European Economic Integration

Edited by
Christophe Deissenberg, Robert F. Owen
and D

Published as a special supplement to
Review of International Economics
Volume 5 Issue 4

Copyright © Blackwell Publishers Ltd 1997

First published in 1997

Blackwell Publishers
108 Cowley Road
Oxford OX4 1JF, UK

and

350 Main Street
Malden
MA 02148, USA

British Library Cataloguing in Publication Data

A CIP catalogue record for this book is available from the British Library

Library of Congress Cataloging-in-Publication Data applied for

This book is printed on acid-free paper.
Printed and bound by Galliard Printers Ltd, Great Yarmouth, UK

ISBN 0 631 20855 0
ISSN 0965-7576

Contents

European Economic Integration: An Introductory Overview
Christophe Deissenberg, Robert F. Owen, and David Ulph   1

The Economic Case for Monetary Union in the European Union
Willem H. Buiter   10

Wage Rigidity, Monetary Integration and Fiscal Stabilization
in Europe
Svend E. Hougaard Jensen   36

Monetary and Fiscal Stabilization of Demand Shocks Within Europe
Chris Allsopp, Gareth Davies, Warwick McKibbin, and David Vines   55

European Nominal and Real Convergence: Joint Process or
Rival Dynamics?
Martine Carré   77

Employment and Wage Bargaining in an Open Monetary Union
Pierre Cahuc and Hubert Kempf   92

The Benefits of Environmental Fiscal Reforms in an
Integrated Europe
Carlo Carraro and Marzio Galeotti   111

Regionalism and the Rest of the World: Theory and Estimates
of the Effects of European Integration
L. Alan Winters   134

Review of International Economics

Review of International Economics, Special Supplement, 1–9, 1997

European Economic Integration:
An Introductory Overview

Christophe Deissenberg, Robert F. Owen, and David Ulph

1. Introduction

In many respects both the overall process of European economic integration and the economic performance of the European Union appear to stand at crossroads. On the one hand, the heightened economic integration following the Single Europe Act, the "1992 Program" and the Treaty of Maastricht offers a unique paradigm of extensive regional market liberalization and international coordination of economic policy making. On the other hand, unemployment and slow economic growth seem to have confounded certain of the initial assessments of the effects of European economic integration, while dampening some of the enthusiasm for continued European policy initiatives aimed at new dimensions of integration and economic cooperation. At the same time, both the process of, and calendar for, European Monetary Unification entail immediate performance implications and potential policy tradeoffs which will undoubtedly add to the complexity of the analysis and debates associated with European economic integration. Given both the economic size of the European Union and the novelty of many facets of the integration experiment, it is apparent that the economic issues surrounding the European integration process are of global significance.

While there has been an extensive amount of recent research on a variety of specific analytical, empirical, and policy issues regarding European economic integration, it has become increasingly apparent that many of the questions invoked are far more complex and fundamental than was initially perceived. They include both the location and transformation of economic activity, as well as the design and coordination of economic policy associated with the process of economic integration. It is clear that such issues are of archetypal significance for understanding not just contemporary developments within Europe and their relation with the world economy, but also similar phenomena worldwide. These will become of increasing importance in the face of not only heightened integration within different economic regions, but also an increasingly globalized world economy, which is characterized by reduced international transaction costs and growing economic interdependence.

The aim of this special volume of the *Review of International Economics* is to highlight fundamental conceptual, empirical, and policy issues that underlie a range of current European economic concerns, while also underscoring their wider global significance. An initial set of three papers examine several different aspects of the potential macroeconomic consequences of the envisaged European Monetary Union (EMU). Specific aspects of the analysis include an evaluation of the case for and against EMU, as well as the consequences of monetary unification for the optimal design of macroeconomic policies in the European Union (EU). A related contribution then offers an assessment of the extent to which European economies can be characterized by both financial and real convergence. The three final papers are more focused on real-side implications of economic integration. An initial paper emphasizes how a combination of labor market segmentation and differences in wage-setting

paradigms can influence wage-employment outcomes when there is monetary unification. There is a subsequent analysis of how the effects of taxing pollution are interrelated with the process of economic integration and the extent to which governments
coordinate tax policies. A final contribution offers an assessment of the effects of
European integration on the rest of the world.

The papers in this issue serve both to draw out certain of the lessons that emerge
from past research and to delineate the state of current research on a range of topics
related to European economic integration. They also identify a series of unanswered
questions thereby defining the frontier for future research on a subset of major issues
arising from European economic integration.

2. Overview of the Specific Paper Contributions

The prospect of European Monetary Union marks a potentially new and distinctive
stage in the political–economic experiment of European integration. At the same
time that proponents of monetary integration, including notably the European Commission, have highlighted advantages of a single currency, there have also been significant policy concerns. In part, these result from a perception that the fiscal
and monetary policy options open to European nations are rapidly becoming more
limited. This comes at a time when a number of EU countries are experiencing
record unemployment and slow growth which may, at least in part, be due to their
difficulties in adjusting to macroeconomic shocks. An assessment of the opportunity
cost of the hypothetical loss of macroeconomic flexibility associated with the elimination of possible adjustments in intra-EU exchange rates is clearly crucial to an evaluation of the economic rationale for and against EMU. More generally, it is crucial to
analyze both the potential microeconomic and macroeconomic consequences of monetary unification, along with the institutional and policy-design adjustments it makes
necessary.

In "The Economic Case for Monetary Union in the European Union," Willem
Buiter offers a rigorous critical evaluation of the potential effects of European Monetary Union by reviewing both analytic and policy arguments based on the optimal
currency area literature. In particular, he scrutinizes the gains and losses that a nation
can potentially experience by giving up its monetary sovereignty, as well as the optimal
institutional and policy-design changes that should accompany an associated loss of
sovereignty. Such an examination is essential since the theory of optimal currency area
is, implicitly or explicitly, at the core of much of the seemingly contradictory arguments
advanced in the cases for and against EMU. Yet, in light of an array of rather different
paradigms offered in the existing theoretical analysis of optimal currency areas, this is
a challenging undertaking. Indeed, Buiter does not confine his analysis to a single
modeling framework. Instead, he proposes a broad mix of theoretical and empirical
arguments to address in six main sections a series of distinct but, nonetheless, interrelated topics.

The first section of Buiter's paper is devoted to a critical presentation of the classical
microeconomic efficiency arguments for a common currency, which include issues
related to the transparency of prices and the avoidance of international transaction
costs. In the second section, the author addresses the neoclassical public finance
arguments related to seigniorage and the inflation tax. The third section is concerned
with the implications for monetary unification of short-run nominal rigidities in wage
and/or price setting behavior under diverse kinds of asymmetric shocks. There is also
a presentation of recent empirical evidence regarding the effects of such shocks. In the

fourth section Buiter seeks to clarify the policy, institutional and behavioral changes required to compensate for the loss of the nominal exchange rate instrument.

While the analysis of the four initial sections only suggests a weak case for monetary unification, in the fifth section Buiter maintains that possibly only the fully unequivocal argument for a monetary union is critically related to the role of capital mobility. He concludes that unrestricted international mobility of capital, as it is realized within the EU, makes a common currency at least desirable, and presumably unavoidable. Buiter forcefully supports this statement with arguments which principally focus on the inefficient functioning of exchange rate regimes, and on the practical unsustainability of alternative exchange rate regimes in the presence of full capital mobility. The final section of the paper offers an assessment of the usefulness of the Maastricht convergence criteria as a basis for optimally forming monetary unification for a set of European countries. Although Buiter agrees with the need for an exchange rate criterion as a means for avoiding the risk of strategic last-minute devaluation, he rejects the other criteria either as irrelevant or, in the case of the excessive deficit criterion, as both unnecessary and dangerous. The latter assessment has assumed heightened significance in light of current policy concerns about the deflationary impact of restrictive fiscal policies. Constrained by the 3% target for government budget deficits, a number of EU countries appear to have only limited macroeconomic policy options at a time when they are experiencing continued high levels of unemployment and a protracted stagnation of economic growth.

Buiter's overall conclusion is both clearcut and provocative. He maintains that most of the economic arguments usually advanced for or against the monetary union are both misconceived and overstated. Since there is no strong economic case for or against EMU, the issue of whether to unite or not to undertake monetary unification can be viewed as essentially a political rather than an economic one.

A major issue in the debate about monetary integration within Europe regards the tightness of the constraints on fiscal policy that are entailed by the convergence criteria linked to the Treaty of Maastricht. The concern is that if both monetary and fiscal policies are very tightly constrained then the impact of shocks—particularly asymmetric shocks—may be more severe than if some degree of flexibility in policy is allowed. These concerns are particularly acute if labor markets are also characterized by a high degree of rigidity. Both the paper by Svend E. Hougaard Jensen and the article by Chris Allsopp, Gareth Davies, Warwick McKibbin, and David Vines examine specific aspects of the foregoing general questions.

In his paper, "Wage Rigidity, Monetary Integration and Fiscal Stabilisation in Europe," Jensen explores how the degree of fiscal flexibility that might be required within a monetary union will depend on both the degree of flexibility in the labor market and on how tight are the restrictions on exchange rate movements. He demonstrates a number of interesting results. The first is that even if the labour market is flexible, and so can act as a shock absorber, asymmetric shocks can still threaten macroeconomic stability. Nonetheless, the more flexible is fiscal policy, the greater will be the speed of stabilization for Europe as a whole, and there will be a more rapid convergence in the performance of individual countries. A second finding is that if labor markets are rigid, then flexible fiscal policy can never completely remove the effects of asymmetric supply shocks. Nonetheless, it can successfully contribute to achieving heightened convergence between member states. A third conclusion is that introducing a degree of flexibility in monetary policy does not significantly reduce the need for flexibility in fiscal policy. Indeed, within the framework of Jensen's model, as long as fiscal policy is flexible, monetary union performs well, since it automatically insures against financial shocks which can appear in regimes with monetary flexibility. While this analysis is of principal interest to debates about European integration, the general lessons about the need to take account of the flexibility of the overall institution environment, when considering the appropriate design of optimal macroeconomic policy within an economic and monetary union, are of more general relevance.

In their article "Monetary and Fiscal Stabilisation of Demand Shocks within Europe," Chris Allsopp, Gareth Davies, Warwick McKibbin, and David Vines examine how the effectiveness with which macroeconomic stabilization policies can respond to demand shocks, in a given economic region, depends critically on exchange rate regimes and the extent to which fiscal policies are constrained. While this research shares the general policy concern of other papers in this special issue, regarding the possible macroeconomic constraints implied by EMU, the authors' approach is quite distinctive. In particular, they juxtapose analytic results based on relatively simple IS–LM–BP models, involving either a single country or a two-country plus rest-of-the-world framework, with simulations using the large and sophisticated, global economic model developed by McKibbin and Sachs. The latter model, denoted MSG2, is a multiregional world dynamic model which has incorporated intertemporal optimization of agents' decisions in a framework which emphasizes the role of microeconomic foundations for the specification of the model's structural equations.

The initial theoretical modeling underscores how the alternative scenarios of floating exchange rates and monetary unification have quite different macroeconomic implications for an economic region, depending on whether demand shocks are positively or negatively correlated between countries. Under a monetary union, while a given positive demand shock in one country will translate, *ceteris paribus*, into a source of increased export demand from a partner country, there will be an offsetting effect due to an appreciation of the region's single currency in relation to a representative currency in the rest of the world. Clearly, the existence of built-in fiscal stabilizers will attenuate both the domestic and intraregional effects of the initial demand shock. Accordingly, as in the subsequent simulation exercises, a major focus is on how the consequences of monetary unification for macroeconomic performance depend on alternative fiscal scenarios. Specifically, three cases are considered in which tax rates either are (1) set to zero, so that fiscal stabilizers are inoperative, (2) equal an exogenous rate, or (3) are allowed to adjust to compensate for changes in GDP.

In the simulations, a European version of the MSG2 model is then used to assess the effects of asymmetric demand shocks originating in a single country under different monetary and fiscal regimes which highlight the potential implications of EMU. More specifically, the two principal simulations undertaken relate to (1) a 4-year consumption shock originating in the United Kingdom, which is understood to reflect the boom performance of the British economy in the late 1980s, and (2) the effects of German Economic and Monetary Union (GEMU). The simulation results, which include comparisons between a floating regime where inbuilt stabilizers are working and fixed exchange rate regimes for the three different fiscal cases, are found to be quite consistent with the initial theoretical insights derived by the authors.

The overall analysis in this paper highlights two principal concerns regarding possible consequences of EMU for macroeconomic stabilization policies. The first involves the possibility that EMU could lead to excessive central monitoring and control of fiscal policies, thereby weakening the effects of fiscal stabilizers operating within each member country. A second fear is related to the so-called "Kenen problem," whereby overall monetary policy in Europe might be overly designed to respond to specific demand shocks in just one country/region, such as that which arose under GEMU. In light of the shared recent macroeconomic slowdowns of a number of countries in the EU, it is clear that there appear to be substantial heuristic grounds for both of the authors' concerns. In particular, the authors' analysis underscores certain of the economic dangers of an overly strict application of the deficit rules indicated by the Treaty of Maastrict in the current high-unemployment/low-growth environment characterizing many EU economies. This latter conclusion echos the previously discussed findings of Buiter and Jensen.

The perceived need to let only countries with similar economic structures and performance enter the planned EMU led to the definition of a number of nominal convergence criteria as prerequisites for EMU membership. As set forth under the Treaty of Maastricht, these criteria relate to levels of inflation, interest rates, stocks and flows of governmental debt, as well as limitations on exchange rate adjustments. However, it remains open to question whether nominal convergence along such lines will actually entail real convergence and, thereby, have the desired structural consequences on economies within the Union. Investigating the link between both types of convergence is therefore crucial for evaluating the appropriateness of the financial criteria set forth by the Treaty.

Martine Carré's article, "European Nominal and Real Convergence: Joint Process or Rival Dynamics?", offers a valuable perspective on the links between real and financal performance in the EU. The aim of the proposed analysis is to empirically evaluate, using existing data for the European Community and EFTA countries, over which subintervals of the period 1964–1992 the real and nominal processes have been positively or negatively correlated. Carré relies on the standard concepts of sigma-convergence and beta-convergence, which have been used in previous empirical studies of the convergence between European countries. β-convergence measures the tendency for a poor economy to grow faster than a rich one, while σ-convergence relates in this instance to the decrease over time of the cross-variance of per capita GNP. The originality of the article is to simultaneously study nominal and real convergence by, respectively, using the proxies of the inflation rate and of per-capita social product. A joint system of equations is estimated to test the sensitivity of the convergence process with respect to the initial per capita GNP, and to make explicit the interrelation between nominal and real convergence, which is shown to be changing over time. In particular, Carré finds that the nominal convergence of the early 1980s was accompanied by a real divergence. This can presumably be attributed to the deflationary policies widely adopted after the creation of the European Monetary System. However, over the 1987–1992 period there appears to have been simultaneous convergence in the financial and real measures. While such evidence could be interpreted as lending strong support for the viability of European monetary unification, there remains the further question of the overall optimality of the observed convergence paths and the underlying links between financial and real performance.

One of the major issues that arises in the debate over a single European currency concerns the potential internal adjustments that may arise between countries in a monetary union when the possibility of internal exchange rate adjustments is ruled out. While an analysis of this question is relatively straightforward when all internal and external real and financial markets are perfectly competitive, it is inherently much more complex when markets are imperfectly competitive. In a European context the case of imperfectly competitive labour markets is of particular relevance, so that there is the perspective of a monetary union where national labor unions may play quite different roles in the determination of wage rates and employment in individual economies. In the absence of any exchange rate adjustment, there would consequently be significant interactions between the wage rates set in one country, and, via product markets, the labour market conditions in other countries. Intuitively, what this implies for the equilibrium levels of wages and employment depends on (1) how important these interactions are to workers—which depends, in turn, on the openness of the union to trade with the rest of the world; (2) the degree of substitutability between the goods produced by the various countries in the monetary union; and (3) the nature of the bargaining processes within the union, since these will determine the extent to which wage settlements in one country may be influenced by their impact on the labour market performance of other countries in the monetary union.

The paper by Pierre Cahuc and Hubert Kempf, entitled "Employment and Wage Bargaining in an Open Monetary Union," explores formally how the foregoing factors combine to affect the equilibrium wage/employment outcomes. They consider a simple three-country model, with two countries in monetary union facing a large third country—the rest of the world. In each of the first two countries there is a separate union which sets wages. The crucial feature is that wages have to be set for two periods. This allows them to consider three types of wage-setting regime. In the first, wages are set both independently and synchronously in each country, so essentially each union is assumed to act in a standard Nash fashion, while ignoring the effect of its behavior on the other union. In the second regime, wages are again set independently in each country, but now there is staggered wage-setting. Each union takes account of the effect the wage it sets for this period will have on the wage set by the other union in the next period, such that there is Stackelberg behavior. Finally, Cahuc and Kempf consider centralized bargaining where unions act cooperatively. They show that the ranking of wages and employment across these three types of bargaining arrangements depend on just two parameters. The first of these is the degree of product market substitution between each good produced in one of the two countries in the monetary union. The second is the overall weight in consumers' utility function that is attached to these two products—as distinct to a representative import good from the rest of the world. This latter parameter reflects the degree of openness of the monetary union to trade with other countries outside the union. Since these two parameters determine both the nature of the interaction between the two labor markets and the extent to which unions will care about such interdependence, it is not surprising that this will affect the ranking of wage–employment outcomes across the three bargaining regimes.

The foregoing results provide important insights on issues related to the prospective single European currency. However, the theoretical insights of this paper are not confined to a European setting, and apply, more generally, to other monetary unions. Yet, as Cahuc and Kempf admit, there are certain special features of their model which need to be examined further. For example, the countries in the monetary union are assumed to be symmetric. Given the significance and generality of the issues addressed in this paper, it would be most worthwhile to investigate further the novel approach proposed by the authors.

One of the more interesting and controversial ideas in the recent environmental literature has been the "double dividend hypothesis." Stated loosely, the idea here is that if a government increases (nondistortionary) pollution taxes this will bring about two potential benefits. First, the level of undesirable activities, such as pollution, will be reduced. Second, the revenue generated from such taxation so raised will enable the government to cut other, more distortionary taxes, thereby expanding desirable economic activities. In particular, in economies experiencing involuntary unemployment, there is the prospect of significant job creation. While there have been a number of different theoretical formulations of this idea, the general consensus that seems to emerge is that while such a double dividend is certainly theoretically possible, it is by no means inevitable. Indeed, rather strict conditions need to be satisfied for it to arise. Furthermore, theoretical work indicates that the functioning of labor markets is crucial to the existence of certain forms of the double-dividend scenario. These theoretical insights are supported by a corresponding mixture of empirical findings on the effects of increased pollution taxes. However, the existing focus has been on a context in which there is a single government. When extended to a multi-jurisdictional context—such as Europe—new dimensions of complexity are added to the question. In particular it is important to consider how far the possibility of there being a double dividend now depends on the extent to which pollution policies and/or subsequent fiscal reforms are coordinated internationally.

It is to this issue that the paper by Carlo Carraro and Marzio Galeotti, "On the Benefits of Environmental Fiscal Reforms in an Integrated Europe," is devoted. Their analysis benefits from the use of a new quantitative general equilibrium multiregional model of the EU, which allows a sophisticated modeling of the labor market in each country while also capturing the generation of carbon-dioxide emissions. Three different scenarios are simulated. In the first of these, countries agree on the general design of the overall tax package—which pollution taxes may be raised, which other taxes are to be cut—but each country is free to decide the specific amounts by which taxes are to be raised or lowered. This is the completely decentralised scenario in which the subsidiarity principle is fully respected. In the second setting, countries agree on the specifics of the pollution tax reform, and so impose a uniform carbon tax, but leave individual countries free to decide how to use the associated revenue. The final scenario corresponds to a "federal" solution, where there is full coordination of both aspects of the policy package.

The main findings are as follows. First, while, in the short run, these reforms bring about some sharp reductions in aggregate CO_2 emissions and some modest increases in aggregate employment, in the long run the effects are very small. In part this is due to the induced growth of output and employment which raises pollution levels. Second, reducing payroll taxes is a very indirect and rather ineffective way of liberalizing the labor market. Third, the strongest long-run effects on employment arise in the "federal" case. Yet, precisely because of this expansionary effect, the long-run pollution reduction is smaller than in the case where countries just harmonize the carbon tax.

At the same time that the paper by Carraro and Galeotti is immediately pertinent to European policy debates, their analysis of the implications of being able to achieve coordination of international pollution and/or fiscal policies is also of much wider relevance.

The repercussions of European integration for economic performance in the rest of the world (RoW) are assessed in L. Alan Winters' paper, which is entitled "Regionalism and the Rest of the World: Theory and Estimates of the Effects of European Integration." One of the reasons why such a study of the international spillover effects of European integration is important is because it is just one of a number of different examples of regional integration taking place throughout the world. From a global perspective it is far from clear whether such exercises in regional liberalization are to be encouraged or discouraged. An essential step towards formulating such a policy evaluation is to assess the effects of regional integration on economic welfare outside a given area.

Winters starts by noting that existing *ex post* estimates of the effects of integration are flawed because they attempt to draw welfare conclusions based on the analysis of trade flows between the integrating region and the RoW. In general, this will give a very incomplete and inadequate indication of the true welfare effects. Winters therefore begins his analysis by adapting the methodology proposed by Baldwin and Venables for measuring the welfare effects of integration to the case where the focus is on the effects of integration on the RoW. He shows that there are three sets of factors that need to be taken into account: (1) static perfectly competitive effects; (2) static imperfectly competitive effects; and (3) the dynamic accumulation effect. In the first instance, static perfectly competitive effects arise through: (a) changes in trade volumes—though these are relevant only to the extent that there are trade distortions, and the sign and magnitude of this effect depends on the precise nature of these distortions; (b) changes in trade costs induced by integration; and (c) changes in the terms of trade. Effects arising from imperfect competition arise to the extent that industries deviate from the textbook model of perfect competition and comprise (a) output expansion effects (resulting in reductions in calculated deadweight losses, based on the Harberger method); (b) scale effects as consumers gain from the reduction in prices that arise as firms exploit economies of scale; and (c) variety effects as the diversity of products available to consumers in RoW changes through integration. Finally, the accumulation effects arise whenever integration generates changes in investment and there is a deviation between the discount rate and the social rate of return.

Within such a more general, conceptual framework it is not possible to get *ex post* measures of the welfare effects, while there are also major obstacles to determining *ex ante* estimates. The first of these is that considerable judgment is needed in order to obtain an appropriate specification of the precise comparative static effects resulting from integration. Often the effects are model-specific and there may not be valid theoretical grounds for even signing certain of them. Second, as already noted, certain effects depend on the signs and magnitudes of pre-existing distortions, and there can be considerable problems in their assessment. Finally, there are also sheer data problems inherent to the relatively complex calculations.

Winters offers a quite rigorous analysis which makes a serious attempt to quantify the RoW effects by surveying a number of existing studies. However, since much of the existing research does not consider the RoW explicitly, often only indirect inferences can be drawn. The conclusion is that the RoW probably does lose from a regional integration such as that in Europe, though the impact on economic welfare is likely to be small. These conclusions are admittedly tentative, and there is clearly scope for developing the work further. However, the most important contributions of the paper are to demonstrate not only the generality of the generic welfare calculations aimed at assessing the global consequences of European economic integration, but also to show just how difficult it is to draw firm conclusions about such welfare estimates.

3. Acknowledgement

The papers appearing in this special issue were initially presented at an international conference on "European Economic Integration," held 8–9 June 1995 in Nantes, France, under the patronage of the President of the European Commission, Jacques Santer. This conference on "European Economic Integration" was jointly organized by the "C3E" Research Center at the Faculty of Economics and Business Administration at the University of Nantes and by the French Economics Association ("AFSE"). The guest editors of this volume wish to thank Patrick Artus, Jean-Yves Caro, Lionel Fontagné, and Pierre-Yves Hénin for their contributions as other members of the conference organization committee. Gratefully acknowledged also is the assistance of members of a scientific committee, which comprised Jean-Claude Berthelemy, Olivier Blanchard, Paul Champsaur, Jacques Crémer, Jacques Drèze, Barry Eichengreen, David Encaoua, Jean-Pierre Gourlaouen, Paul de Grauwe, Alexander Italianer, Alexis Jacquemin, Pierre Mohnen, Pierre Morin, Jean-Louis Mucchielli, Pierre-Alain Muet, Jean Pisani-Ferry, Jean Tirole, David Vines, Alan Winters, and Charles Wyplosz.

The organization of the conference was made possible by generous financial support from the European Commission, the French Education and Research Ministry, the Central Bank of France, the Commissariat Général au Plan, the Caisse des Dépôts et Consignations, the Crédit Agricole, the Office of the Mayor of Nantes, the Nantes Chamber of Commerce, the Conseil Général of Loire–Atlantique, and the Department of Economics at University College London. The financial contributions of the Scientific Council, the Institut Universitaire Professionalisé Banque-Finances Europe, and the Department of Economics of the University of Nantes are also gratefully recognized. Finally, we are grateful to the many anonymous referees who reviewed papers presented at the conference, and whose comments were critical to the selection of the papers to appear in this special issue.

Review of International Economics, Special Supplement, 10–35, 1997

The Economic Case for Monetary Union in the European Union

Willem H. Buiter*

Abstract

Differential requirements for seigniorage provide a weak case for retaining monetary independence. As regards adjustment to asymmetric shocks, nominal exchange rate flexibility is at best a limited blessing and at worst a limited curse. Absence of significant fiscal redistribution mechanisms among EU members is not an obstacle to monetary union. Neither is limited international labour mobility. Convergence of real economic performance is irrelevant for monetary union. A common currency is the logical implication of unrestricted capital mobility. The Maastricht criteria need not hinder monetary union provided the political will exists to adopt a flexible interpretation of the fiscal criteria.

1. Introduction

This paper addresses the question: "What economic considerations should determine whether the countries belonging to the European Union ought to be pursuing monetary union," or "Is the EU an Optimal Currency Area?" I approach the issue from a narrow economic perspective. Most of the advocates and proponents of monetary union in the EU are not motivated by such economic considerations. Monetary union is first and foremost the next step in an ongoing process of economic and political integration in Western Europe.

The focus is throughout on the policy implications of the various strands of the optimal currency area literature: what does a nation (or group of nations) gain or lose by giving up monetary sovereignty, and what changes are required (in the fiscal policy instrumentarium and/or in other institutional or policy design aspects of the capacity to adjust) to compensate for the loss of the national monetary instruments? Unfortunately, the theory of optimal currency areas is one of the murkiest and most unsatisfactory areas of macroeconomic and monetary theory.

The outline of the paper is as follows. Section 2 reviews microeconomic arguments for a common currency. Section 3 reviews neoclassical public finance arguments against a common currency. Section 4 considers the implications of nominal rigidities. Section 5 looks at changes in policies and institutions required to make up for the loss of the exchange rate instrument. Section 6 reviews the role of capital controls and Section 7 the Maastricht convergence criteria.

---

*Buiter: University of Cambridge, Sidgwick Avenue, Cambridge CB3 9DD, UK. Tel: 00 44 (1223) 335210; Fax: 00 44 (1223) 335475; Email: willem.buiter@econ.cam.ac.uk. This contribution is based on my paper "Politique Macroéconomique dans la Période de Transition vers l'Union Monétaire" (Buiter, 1995) and on Chapter 10 of *Financial Markets and International Monetary Cooperation: The lessons of the 92–93 ERM Crisis*, written with Giancarlo Corsetti and Paolo Pesenti. Apart from my two coauthors, I am indebted to Niels Thygesen, Arvind Panagariya, Georg von Furstenberg, Jacques Melitz, Pierre-Yves Henin, Yves Herve, Mike Wickens, Patrick Minford, Marcus Miller, Martin Weale and David Vines, for helpful comments on earlier versions of the material covered. An anonymous referee provided comments and suggestions for improvements that were both detailed and wide-ranging. Financial support from ESRC grant R02220032, "The Ins and Outs of Staggered Economic and Monetary Union in Europe," is gratefully acknowledged.

2. Microeconomic Efficiency Arguments for a Common Currency

The microeconomic efficiency arguments for a common currency are well-known. A medium of exchange or transactions medium is subject to an obvious *network externality* (Dowd and Greenaway, 1993). This is most easily seen in the case of intrinsically valueless (or fiat) money: the usefulness of a medium of exchange and the likelihood of it being accepted in exchange for intrinsically valued goods and services by an economic agent is increasing in the number of other economic agents that are likely to accept it as a medium of exchange, since what determines the *liquidity* or *moneyness* of the medium of exchange is the probability of being able to dispose of it when desired, at short notice, and at a low and certain cost.

The use of a given stock of money balances in transactions is obviously *rival*: I can only spend a given dollar bill once. However, since the usefulness to me of any particular currency for effecting transactions is strictly increasing in the frequency, scale, and scope of that currency's use by others, there is an *ultra-nonrivalness* in the choice of which currency to use. This creates the public good aspects of money.[1] Social transactions costs are minimized with a single currency. There is a direct parallel here with the social gains from having a common language (or a common measurement system):[2] apart from aesthetic considerations, the value to me of learning another language is increasing in the number of other people that know the language. The benefits from having a common currency are a continuing flow of real resource savings, now and in the future.

Complex measurement problems make it difficult to assess empirically the order of magnitude of the microeconomic efficiency gains that might be achieved by monetary union, in the EU or elsewhere. The Cecchini Report tried to estimate the real resource savings from the bid–ask spreads in the foreign exchange markets. This, the value added in the foreign exchange business, represents the competitive rentals of the physical and human resources currently tied up in the exchange of currencies that would be liberated by monetary union, plus any pure rents, enjoyed either as monopoly profits or as X-inefficiency (organizational slack). To the extent that these markets are imperfectly competitive, the equilibrium spreads overstate the social opportunity costs incurred by banks and other foreign exchange traders, of exchanging one currency for another. On the other hand, the spread ignores altogether the real resource costs incurred by the other (non-bank) parties in the foreign exchange markets, the so-called *inhouse costs* in Emerson et al. (1990). In addition, a whole range of further securities (bonds denominated in the national currencies, options, futures contracts and other derivatives) would become redundant following monetary union, freeing additional resources for alternative uses. In sum, the existing estimates are incomplete, inconclusive, and rather arbitrary, and there is no consensus regarding the true magnitude of the net microeconomic benefits from having a common currency.

While, if one could redesign the world from scratch, microeconomic efficiency would clearly suggest the optimality of a single common currency, it does not follow that it is necessarily efficient to move to a common currency from an initial situation involving many currencies. *Switching* currencies is costly in a world of boundedly rational agents with limited computational, data-gathering and data-processing capacity.[3]

In addition, there are the real resource costs of introducing a new currency (or of extending the use of an existing currency to previous nonusers), the costs of converting contracts denominated in old currencies into the new currency, and a variety of other costs that can be labelled "vending machine costs." To sum up, the *one-off* cost of switching must be set against the continuing gains from operating with a single currency, meaning that the microeconomic case for moving to a common currency from a pre-existing multiple currency system is not *a priori* self-evident.

An interesting point, noted in Dowd and Greenaway (1993), is that, from the point of view of efficiency, a move to a single currency should be a move towards the universal use of one of the pre-existing currencies (say the D-mark in the EU) rather than the adoption of a new currency (such as the euro). That way at least the Germans will be spared the switching costs, and even non-Germans will be dealing with a common currency that will at least be somewhat familiar. By the same token, English, Spanish or Mandarin would make a better world language than Esperanto. This suggests the following (somewhat tongue-in cheek) proposition.

PROPOSITION 1. *If the European Union (or a subset thereof) moves to a common currency, efficiency considerations suggest that the name of the most widely used existing currency be attached to the new common* numéraire. *This means that the name of the new European currency should be the D-mark.*

Nothing in the Maastricht Treaty precludes the adoption of the D-mark as the *name* of the common currency. Note again that while the name "D-mark" would be retained, the Bundesbank would, as provided in the Maastricht Treaty, lose its ability to conduct monetary policy in Germany or anywhere, and would become just the German branch office of the ECB. It would be the ECB that controlled the EMU-wide issuance of D-marks following monetary union.[4]

Leaving aside the microeconomic efficiency arguments for a common currency, there are just two reasons why the nominal exchange rate regime might matter for real economic performance: *seigniorage* and *nominal inertia*. Note that, apart from the microeconomic efficiency arguments, the arguments for a common currency are the same as those for any *credible* fixed exchange rate regime. It may of course be the case that the only truly credible fixed exchange rate regime is a common currency.

3. Exchange Rate Independence and Seigniorage

Basic Concepts

Governments[5] can appropriate real resources by issuing intrinsically valueless (fiat) money, provided private agents believe that fiat money will offer them a competitive rate of return (including saved transactions costs) over the planned holding period.

Let the nominal quantity of government fiat money (henceforth base money) outstanding at the end of period t be denoted H_t. For simplicity, assume that base money (currency plus banks' balances with the central bank) is non-interest-bearing. Let P_t be the general price level during period t, and Y_t be the real GDP; Δ is the backward difference operator. It is important to avoid loose language and to distinguish between "seigniorage" and the "inflation tax." Seigniorage is the value of the resources the government appropriates by expanding the nominal monetary base. As a fraction of GDP, it is given by σ_t:

$$\sigma_t \equiv \Delta H_t / (P_t Y_t). \tag{1}$$

There is a closely related concept, occasionally also referred to in the literature as seigniorage (although I shall avoid that usage), given in equation (2), which defines the interest burden foregone by the government through its ability to issue non-interest-bearing liabilities. Let i_t denote the one-period nominal interest rate on government interest-bearing debt issued in period t. This concept of interest burden foregone (the *opportunity cost* measure of seigniorage), denoted ω_t, is (as a fraction of GDP):

$$\omega_t \equiv i_t H_{t-1}/(P_t Y_t) \equiv i_t h_{t-1}/[(1+\pi_t)(1+g_t)] \tag{2}$$

where $h_t \equiv H_t/(P_t Y_t)$, $1+\pi_t \equiv P_t/P_{t-1}$, and $1+g_t \equiv Y_t/Y_{t-1}$.

The flows of current and future seigniorage and the flows of current and future interest burden foregone are related by the following identity:

$$\sum_{j=0}^{\infty}\left(\prod_{k=0}^{j}(1+i_{t+k})^{-1}\right)\Delta H_{t+j} = \sum_{j=1}^{\infty}\left(\prod_{k=0}^{j}(1+i_{t+k})^{-1}\right)i_{t+j}H_{t+j-1} - H_{t-1}/(1+i_t). \tag{3}$$

Thus, the present discounted value of current and future seigniorage equals the present discounted value of the current and future interest burden foregone minus the initial stock of base money (the liabilities of the central bank).

A third related concept, also at times referred to as seigniorage, is the Central Bank's budgetary contribution to the general government. This is effectively the tax levied by the Treasury on the central bank. For our purposes, the *intra-public sector* transfer of resources between the central bank and the general government is of no interest.[6] What matters is the transfer of resources between the public sector as a whole (that is, the consolidated general government and central bank) and the remaining economic actors (the domestic private sector, the state enterprise sector, and the rest of the world).

The *inflation tax* is generally defined as the reduction in the real value of the outstanding stock of base money due to increases in the general price level. Thus, the inflation tax in period t, as a fraction of GDP, τ_t^π, is given by

$$\tau_t^\pi \equiv \pi_t H_{t-1}/(P_t Y_t) \equiv \pi_t(1+\pi_t)^{-1}(1+g_t)^{-1}h_{t-1}. \tag{4}$$

The inflation tax and seigniorage are related by the identity

$$\sigma_t \equiv [(1+\pi_t)(1+g_t)-1](1+\pi_t)^{-1}(1+g_t)^{-1}h_{t-1} + \Delta h_t$$
$$\equiv \tau_t^\pi + g_t(1+g_t)^{-1}h_{t-1} + \Delta h_t. \tag{5}$$

Seigniorage equals the inflation tax plus the "real growth bonus" $g_t(1+g_t)^{-1}h_{t-1}$, plus the increase in the monetary base-GDP ratio Δh_t.

When the demand for money is sensitive to the (expected) rate of inflation, the inflation tax is distortionary, like every other real-world tax, transfer, or subsidy. The normative neoclassical theory of public finance recognizes that, in general, a (con-strained) optimal design of fiscal policy will require the use of all distortionary tax instruments. Efficiency requires that the excess burdens imposed by the various distortionary taxes be equalized at the margin. This might seem to create a presumption that countries with well-developed direct and indirect tax systems should make less use of the inflation tax than countries with less efficient revenue administrations and more relaxed public attitudes towards tax evasion. The optimal inflation rate might be expected to vary across time and across countries as tax bases, tax administration capacities, and tax ethics vary. This would constitute an argument against a common currency. In addition, the inflation tax is one of the few means of taxing the (cash-intensive) underground economy, which may be desirable both on efficiency and on equity grounds (Canzoneri and Rogers, 1990). The importance of this motive may well vary among the current members of the EU.

This presumption is less robust than one might assume, however, even as a purely theoretical proposition. Recent insights into the optimal use of distortionary taxes on the returns from durable (capital) assets, due to Chamley (1986) (see also Lucas, 1990; Corsetti, 1992; Corsetti and Pesenti, 1992; Zhu, 1992; and Roubini and Milesi-Ferretti, 1994) imply that, at least in the fairly standard theoretical model developed in Buiter (1995), the Friedman rule for the optimal quantity of money (the nominal rate of interest should be zero and satiation with real money balances should occur) still applies.

Few people are likely to lie awake about seigniorage for most EU countries in any case. In recent years there has been very little recourse to the anticipated inflation tax or to seigniorage for most EMU countries, with the notable exceptions of Spain, Italy, and especially Greece and Portugal (Grilli, 1989a, 1989b). Without committing the offense of measuring the amount of damage done by a tax (or by its abolition) by the revenue it raises, it seems extremely unlikely that the imposition of a common (low) rate of inflation on the EMU countries would significantly increase the excess burden associated with the financing of the public spending program.

A Broader View of the Inflation Tax

The inflation tax referred to in the theory of public finance is perhaps more accurately referred to as the (narrowly defined) *anticipated* inflation tax. Anticipated inflation can influence the government's budgetary position through other channels. The most important of these is the Olivera–Tanzi effect through which a higher rate of inflation erodes the real value of taxes paid in arrears. The reason is that such arrears often are not index-linked and are not subject to a market interest rate reflecting anticipated inflation.

In addition to using the anticipated inflation tax (broadly defined), the government can improve its real financial net worth by reducing the real value of its outstanding nominally-denominated fixed interest rate debt through unanticipated inflation. Variable-interest-rate, short-maturity debt can have its real value eroded by an unanticipated increase in the price *level*. Even if nominal domestic costs are sticky, the CPI will be flexible in an open economy through the import component of the consumption bundle. In a small open economy, a price level jump can be engineered through a discrete (or maxi-) devaluation.

Giving up the ability to have nationally differentiated unanticipated inflation tax levies on the national debt, may be more serious than the loss of the discretionary use of the anticipated inflation tax for a number of countries with high public debt GDP ratios and a doubtful capacity for generating significant and sustained primary government surpluses. For this group of countries, which includes Greece, Italy, and Belgium, the option of a *de jure* (through a partial "consolidation" or default by some other name) or *de facto* (through an inflation surprise or an unexpected devaluation) capital levy on the public debt is valuable. If a *de jure* public debt repudiation turns out to be politically unacceptable, a fierce burst of monetary and exchange rate irresponsibility may be the only way to reimpose *ex post* consistency on the public accounts. The optimal time to do this would be just before joining EMU, since in that case there would be no cost (in terms of the credibility of the country's commitment to future noninflationary policies) from having a last fling with inflation.

As the EU is only a relatively small subset of the set of all nations, there is an additional international seigniorage dimension. Member currencies (especially the D-mark) are used as reserves, intervention currencies and vehicle currencies by official and private agents outside the EU. The total amount of external seigniorage raised by all EU members from non-EU members is likely to change as a result of monetary union. It is quite possible that a new European currency could become, in relatively short order, a more effective competitor for the US dollar as an international store of value than the DM is today.[7] This good news must, however, be balanced by the recognition that the rules that will be followed by the European Central Bank for the distribution of its seigniorage (including its external seigniorage) among the various member states are unlikely to mimic the current distribution of seigniorage. Scope for conflict is clearly present.

4. Nominal Rigidities and the Keynesian Arguments for an Optimal Currency Area

The monetary non-neutralities I will focus on in this section are short-run "Keynesian" non-neutralities, due to nominal rigidities in wage and/or price setting behavior.[8] It is ironic that, ignoring the microeconomic efficiency argument and the neoclassical public finance arguments, one can only get legitimately exercised about the abandonment of the national monetary instrument if one believes the economy to have Keynesian features, at least in the short run. Nominal wage and price rigidities are the result of the common practice of setting wages and prices in money terms for months, quarters or even years in advance. These multiperiod nominal contracts are incomplete. In particular, they often are not contingent on nominal wage and price developments elsewhere in the economy or in the economy as a whole: they are not index-linked.[9]

I shall cast the arguments about nominal inertia in terms of the simplest open-economy expectations-augmented Phillips curve, but many other formalizations are possible (Buiter, 1985; Buiter and Miller, 1985). The first issue that must be settled is whether there is any long-run (steady-state) effect of monetary policy on such real variables as the level of capacity utilization or the rate of unemployment. In the Phillips-curve paradigm, long-run non-neutrality of inflation requires at least one of two phenomena to be present: either the long-run Phillips curve is nonvertical or there is hysteresis in the natural rate of unemployment.

The Long-Run Neutrality and Superneutrality of Money

The argument is familiar, so I will restate it only briefly in the simplest possible setting. The actual unemployment rate is denoted u and the natural rate of unemployment u^N. The core inflation rate or underlying rate of inflation is denoted $\hat{\pi}$. The coefficient β measures the weight of foreign prices in the domestic price index; π^* denotes foreign inflation and ε the percentage rate of depreciation of the nominal exchange rate. E_{t-1} is the expectation operator conditional on information at time $t-1$, and ζ denotes some exogenous process driving the natural rate of unemployment. Specifically, ζ is a process independent of past, current, and anticipated future values of the rate of inflation, the growth rate of nominal money or the actual unemployment rate.

$$\pi_t = -\alpha\left(u_t - u_t^H\right) + \gamma\hat{\pi}_t - \beta\left[\hat{\pi}_t - \left(\varepsilon_t + \pi_t^*\right)\right], \quad \alpha > 0, \quad \beta \geq 0, \quad 0 \leq \gamma \leq 1, \tag{6}$$

$$\hat{\pi}_t = \eta E_{t-1}\pi_t + \left(1 - \eta\right)\pi_{t-1}, \quad 0 \leq \eta \leq 1, \tag{7}$$

$$u_t^N = \delta u_{t-1} + \left(1 - \delta\right)u_{t-1}^N + \zeta_t, \quad 0 \leq \delta \leq 1. \tag{8}$$

In a long-run steady state,[10] expectations are realized ($E_{t-1}\pi_t = \pi_t$), the inflation rate is constant, and the terms of trade (or real exchange rate) are constant ($\pi = \varepsilon + \pi^*$). Consider first the case where the natural rate is exogenous, that is $\delta = 0$. For simplicity assume it to be constant as well. In that case,

$$\pi = \alpha\left(\gamma - 1\right)^{-1}\left(u - u^N\right). \tag{9}$$

There is no long-run inflation–unemployment tradeoff if and only if $\gamma = 1$; that is, in the long run core inflation feeds one-for-one into actual inflation and the long-run Phillips curve is vertical at the exogenous natural rate of unemployment.

Now maintain the vertical long-run Phillips curve, that is $\gamma = 1$, but allow *path-dependence* or *hysteresis* in the natural rate by assuming $\delta > 0$. The current natural rate now depends (with exponentially declining weights) on the entire past history of the actual unemployment rate (and, of course, on the entire past history of the exogenous process ζ). While in steady state the Phillips curve is vertical, it can be vertical at any level of the unemployment rate, depending on the past history of the actual unemployment rate. With hysteresis, any temporary shock, including (if there are nominal rigidities) a temporary nominal shock, can have permanent real effects.

The assumption $\gamma < 1$ ceased to be intellectually respectable quite a while ago. The hysteresis hypothesis is intriguing but as yet unsubstantiated. I will therefore maintain the key assumption that neither the nonvertical long-run Phillips curve nor the hysteresis hypothesis are empirically relevant to the EU. This implies that any monetary non-neutralities are strictly short-run.

Short-Run Non-Neutrality of Money and the Implications of Nominal Exchange Rate Flexibility for Real Economic Performance

With money non-neutral in the short run but neutral in the long run ($\gamma = 1$ and $\delta = 0$), both the costs and benefits from nominal exchange rate flexibility are strictly limited and transitory. Nominal exchange rate flexibility makes only a transitory difference to the way in which the real variables of the economy[11] respond to shocks, regardless of whether these shocks are real or nominal and permanent or transitory.[12] The central messages of this subsection are conveniently expressed as a number of propositions.

PROPOSITION 2. *Nominal exchange rate flexibility permits international relative price and cost adjustments that are warranted by fundamental real developments and fundament real shocks—adjustments that will eventually occur regardless of the nature of the nominal exchange rate regime—to be achieved more quickly and with smaller transitional or adjustment costs.*

PROPOSITION 3. *Nominal exchange rate flexibility will cause financial shocks and other nominal shocks to result in temporary changes in international relative*

prices and costs—changes that are unnecessary and harmful from the point of view of
the underlying real fundamentals and that involve real, albeit transitory, adjustment
costs.

PROPOSITION 4. *In a world with incomplete markets, the existence of multiple currencies*
with (potentially) market-determined exchange rates creates additional financial mar-
kets through which extrinsic, nonfundamental or "sunspot" volatility can be injected
into the financial system and thus into the economic system as a whole. Exchange rate
flexibility may breed excess volatility and temporary (but possibly persistent) misalign-
ment rather than merely filtering an exogenously given amount of irreducible, funda-
mental uncertainty.

Asymmetric Shocks

The optimal currency area literature[13] has emphasized that if the preponderance of
shocks hitting a potential common currency area are idiosyncratic or asymmetric—that
is, region-specific or nation-specific shocks—then the case for a common currency is
weakened. Note that the relevant asymmetry can either be asymmetric shocks (im-
pulses) or asymmetric economic structures (domestic transmission, response, or propa-
gation mechanisms). Of course, *nominal* rigidities are a necessary condition for this
conclusion to follow.

Two further characteristics of a country's economic structure have been argued to be
important for the choice of exchange rate regime. These are the openness of the
country to trade in goods and services and the degree of diversification of its production and demand structures.

As regards openness to trade, the argument is that, if importables and exportables
are large relative to domestic absorption and production, then variations in the nominal exchange rate will tend to be translated swiftly and comprehensively into increases
in domestic consumer and producer prices, without any changes in key indices of
international competitiveness. The limiting case would be that of the small open
economy with only traded goods. Note, however, that even in this case nominal wage
rigidity would cause (short-run) changes in real wages and real unit labor costs to result
from variations in the nominal exchange rate.[14] Another weakness of this argument is
that the relationship between openness and the cost of nominal exchange rate rigidity
is obviously nonmonotonic: for a completely closed economy, the nominal exchange
rate regime is a matter of supreme indifference. As regards diversification of production and demand, these are best viewed as determinants of the likelihood that shocks
to the demand for or supply of goods and services are symmetric (general) or asymmetric (nation-specific).

Even in the presence of nominal rigidities, the presumption that asymmetric
shocks favor independent currencies and flexible exchange rates represents at best a
half-truth. This is a simple open economy application of a point first made by Poole
(1970).

Consider a semi-small[15] open economy model with perfect international capital
mobility. All variables are in natural logarithms with the exception of nominal and real
interest rates. Foreign variables and parameters are distinguished by a star superscript.
All parameters are positive. m is the nominal money stock, p the GDP deflator, q the
CPI, s the nominal spot exchange rate (the domestic currency price of foreign exchange), y real output, z the real exchange rate, d the stock of domestic credit, and ρ

the stock of international reserves. The money demand shock,[16] the IS shock, and the supply shock are denoted λ^l_t, λ^d_t, and λ^s_t, respectively.

$$m_t - q_t = k\left(p_t + y_t - q_t\right) - \gamma^l_t + \lambda^l_t, \tag{10}$$

$$y_t = -vr_t + \delta z_t + \lambda^d_t, \tag{11}$$

$$r_t \equiv i_t - E_t q_{t+1} + q_t, \tag{12}$$

$$i_t = i^*_t + E_t s_{t+1} - S_t, \tag{13}$$

$$q_t \equiv (1-\beta)p_t + \beta\left(p^*_t + s_t\right),$$

$$y_t = \alpha\left(p_t - E_{t-1}p_t\right) + \lambda^s_t, \tag{14}$$

$$z_t \equiv s_t + p^*_t - p_t, \tag{15}$$

$$m_t = \theta d_t + (1-\theta)\rho_t. \tag{16}$$

Assume for concreteness that the objective of policy is to stabilize real output around its "full information", natural level λ^s_t.[17] Basically nominal exchange rate flexibility is desirable when faced with "IS" shocks (shocks to the private or public demand for goods and services). For instance, in the face of the German reunification (GEMU) shock, maintaining the nominal exchange rate *vis-à-vis* the D-mark by the other ERM members was bound to be costly for countries with significant nominal wage and price rigidities. Nominal exchange rate flexibility is definitely undesirable in the face of domestic financial market shocks (say liquidity preference (money demand) or shocks to the domestic money supply process). For supply shocks and foreign interest rate shocks the results are qualitatively ambiguous. Without going through a rather tedious full-blown Poole-style analysis, one can still be very precise about the case of monetary shocks.[18]

With a floating exchange rate, $\rho = 0$ (and, for notational simplicity, $\theta = 1$); the money stock, $m = d$, is exogenous. Since our semismall open economy takes the foreign interest rate as given and has perfect international capital mobility, credibly fixing the nominal exchange rate (setting $s_t = E_t s_{t+1} = 0$, say) is equivalent to pegging the domestic nominal interest rate at the level of the foreign nominal interest rate. The now endogenous domestic money stock adjusts passively to shocks in the demand for money through endogenous variations in the stock of international reserves, ρ, even if the stock of domestic credit, d, is exogenous. Real economic activity (output, real exchange rate and real interest rate) is perfectly insulated from domestic financial shocks λ^l_t. So is the domestic price level.

The presumption in favor of interest-rate pegging (of which a fixed exchange rate is the most relevant example for a financially small open economy with perfect international capital mobility) carries over, in the multicountry version of this model, to asymmetric financial shocks.[19] The particular system-wide monetary and exchange rate policy package that is optimal from the point of view of insulating real activity in both countries (and the two price levels), from the effects of monetary shocks, is system-wide nominal interest-rate targeting. The most relevant example of such a policy is a fixed nominal exchange rate, $s_t = E_t s_{t+1} = 0$ (say) (which implies $i = i^*$), and an adjustment of the system-wide quantity of money, $d + d^*$, to keep the common nominal interest rate constant at its target level in the face of monetary shocks in either country or in both countries. Open-loop nominal interest-rate targeting leads to price level indeterminacy in the two-country version of the model: there is no nominal anchor for the system as a whole.[20] The solution to this technical problem is to make some real exogenous variable or policy instrument a function of current, past, or anticipated future values of some nominal price or quantity. An example would be to make the nominal interest rate (the *real* rate of return differential between nominal bonds and base money) a function of the current or lagged price level,[21] e.g.

$$i_t = \hat{\imath} + \eta p_{t-j}, \quad \eta \neq 0; \quad j \geq 0,$$

$$i_t^* = \hat{\imath}^* + \eta^* p_{t-j}, \quad \eta^* \neq 0; \quad j \geq 0.$$

The standard empirical procedure for evaluating the desirability of retaining nominal exchange rate flexibility has involved the decomposition of demand and supply shocks into idiosyncratic or asymmetric versus common or symmetric shocks. Finding a preponderance of asymmetric shocks was then interpreted as an argument against monetary union. Ironically, the standard identifying restrictions commonly imposed in order to distinguish supply shocks from demand shocks imply quite the opposite conclusion. Specifically, the common identifying restriction that demand shocks have no long-run real effects only makes sense for monetary shocks or "LM shocks." The restriction that there be no long-run real effects certainly does not make sense for (permanent) fiscal policy shocks or other "IS shocks."

Empirical evidence (based on credible identifying restrictions) about the relative importance of IS versus LM shocks in the EU is essential for it to be possible to draw sensible inferences about the appropriate exchange rate regime. An equally serious qualification to many of the "shocking" recent findings is that the nature and magnitude of the shocks perturbing the system may be functions of the exchange rate regime itself, as asserted in Proposition 4. That is, different exchange rate regimes not only transmit given fundamental shocks differently, but also may generate different kinds and amounts of extrinsic, nonfundamental noise.[22]

Recent empirical investigations by Maria Nikolakaki (1996) attempt to identify separately the contributions of LM, IS and supply shocks to the variability of output and the real exchange rate in a number of EU countries, using structural VAR methods. In her first identification scheme, only supply shocks can have long-run effects on real output.[23] The aggregate supply shock therefore, by construction, gradually explains most of the output variability across countries as the time horizon increases. She considers nine countries (Austria, France, Italy, Germany, The Netherlands, Portugal, Spain, Sweden, and the UK (all relative to Germany)) for the post-Bretton Woods (1975:4–1994:4) period. Her key findings are as follows. First, supply shocks account for most of the variability in output for Austria, Germany, Sweden, The Netherlands, Portugal, and the UK, even in the short run. In France and Italy, the LM shock typically accounts for between 45% and 68% of output variability over a 1 to 10 quarter horizon. IS shocks account for little of the output variability at any horizon. As regards real exchange rate variability, supply shocks account for very little at any horizon. IS shocks account for the bulk of real exchange rate variability at all horizons except in the Netherlands and Germany, where LM shocks dominate.

In a second identification scheme, both aggregate supply shocks and IS shocks are allowed to have long-run real effects. One implication of this change in identification scheme is that IS shocks become more important as a source of output variability, especially in the long run.

Clearly, even if these results are taken at face value, their interpretation must be informed by the Lucas critique. It is by no means obvious that the reduced-form relationships between output and the real exchange rate on the one hand, and the three shocks on the other hand, will remain invariant under the changes in the stochastic processes driving these three shocks that might result from a major regime change such as a move to a monetary union. Nevertheless, some interest attaches to the finding that LM shocks were a non-negligible source of real output and real exchange rate variability during the post-Bretton Woods era for the nine countries considered.

I summarize this subsection in another proposition.

PROPOSITION 5. *Asymmetric shocks, far from being an argument against a fixed exchange rate or a common currency, are an argument in favor of a fixed exchange rate or a common currency if the shocks in question are financial shocks and the degree of international financial capital mobility is very high.*

5. What is Required to Make Up for Loss of Exchange Rate Flexibility?

What is gained through nominal exchange rate flexibility is an instrument with strictly temporary or transitory real effects. When used properly, it facilitates adjustment to goods market shocks; when used improperly, it may complicate adjustment to financial shocks. Compensating for the loss of the exchange rate instrument therefore requires only an instrument that has strictly temporary or transient real effects.

It is true that the word "temporary" can cover any interval short of eternity. How long is the relevant short run? There obviously can be no answer to this question that is universally valid; it depends on the nature of the shocks hitting the system, on the institutional arrangements in a particular country at any given point in time, and on the decision rules adopted by private agents.

A conventional wisdom going back at least to Milton Friedman holds that in a low-inflation OECD-type economy, rather closed to international trade like the US, it may take as much as two years for monetary changes to feed through into prices rather than affecting real quantities. If capital formation has been affected in the meantime, real consequences of nominal shocks may last longer than that. For more open economies and for economies undergoing higher and more variable rates of inflation, the real consequences of nominal shocks may be significantly less persistent. The UK is probably the European economy with the highest degree of nominal inertia, and even there it is significantly less important than in the USA. There is some evidence to support the view that most of continental Europe has significant real price and cost rigidities, but no nominal inertia of much consequence. The loss of the exchange rate instrument would then be of little importance.

The optimal currency area literature is imprecise and even confused about the policy, institutional or other behavioral changes required to compensate for the loss of the nominal exchange rate instrument. The main confusions concern international factor mobility, international fiscal transfers, and divergent underlying real economic performance.

Factor Mobility

The argument goes as follows. When a country is hit by an asymmetric shock to the demand for or supply of its output, there are two international adjustment mechanisms: first, a change in the relative price of the domestic good, and second, international factor mobility. In a neoclassical world, the two mechanisms are substitutes. If there is a high degree of international factor mobility, international relative prices will have to change only little in response to a given asymmetric goods market shock. Since international relative price changes are costly under a fixed nominal exchange rate, a high degree of international factor mobility, by obviating the need for any (significant) international relative price adjustment, reduces the cost of giving up nominal exchange rate flexibility.

The problem with this argument is that international factor mobility, especially labor mobility, is costly. International relocation of real factors of production is an investment subject to sunk (irreversible) costs. It is therefore efficient only in response to permanent (or at least very persistent) real shocks. Migration flows that are reversible over the typical business cycle are rare. As a substitute for nominal exchange rate flexibility, international factor mobility therefore delivers both too much and too little. It delivers too much insofar as international factor mobility is a mechanism for achieving permanent real adjustments. It delivers too little insofar as it does not possess the self-liquidating, transient properties of a nominal exchange rate adjustment.

It is true that *net* international migration flows can be reversible without this requiring the reversal of any individual migration decision. In a representative agent model, gross flows equal net flows, and reversal of the net migration flow requires individual migrants to reverse their earlier migration decisions; that is, to engage in strictly temporary migration. In a model with a heterogeneous potential migrant population and positive gross flows in both directions, sign reversal in the net flow of migrants between countries does not require any individual migration decision to be reversed; that is, it does not require temporary migration by any individual migrant. By itself, however, recognition of migrant heterogeneity and positive two-way gross flows does not invalidate the presumption that, because migration is subject to sizeable sunk costs, it is neither an effective mechanism for adjusting to temporary shocks, nor an adequate substitute for an adjustment mechanism that has only temporary real effects.

International labor mobility is not an effective cyclical stabilization mechanism. It is a means for achieving long-term structural change. Labor migration is not a very cyclical phenomenon. Since the exchange rate regime affects the behavior of the real economy only at cyclical frequencies, labour mobility is not a substitute for nominal exchange rate flexibility.

The point is often made that the states of the USA are better candidates for a common currency area than the members of the EU, because interstate labor mobility is significantly higher in the USA than intercountry labor mobility in the EU.[24] It is true that in the USA there is rather more permanent or long-term interstate labor mobility, but only little of this occurs at cyclical frequencies. The kind of temporary, reversible, or cyclical international labor mobility required to compensate fully for the loss of monetary autonomy is not found anywhere in the world.

International Fiscal Transfers

What is lost by giving up nominal exchange rate flexibility can be recouped through international fiscal transfers that are strictly temporary or transitory (and indeed reversible—in present-value terms—if there is no Ricardian equivalence).[25] There is no need for any mechanism capable of making permanent fiscal transfers in order to make up for the loss of national monetary autonomy. The fact that the EU budget is tiny and engages in a negligible amount of international redistribution is therefore irrelevant from the point of view of monetary union. By the same token, the fact that the US federal budget is responsible for a significant amount of interstate redistribution represents massive overkill from the point of view of establishing the presumption that the USA is an optimal currency area.[26]

All the EU needs in order to compensate for the loss of national monetary sovereignty and nominal exchange rate flexibility is an international transfer mechanism that is capable of making temporary (i.e., self-liquidating) transfers between countries.[27]

Divergent Real Developments

There is a widely-held view that convergence in real economic performance is a substitute for nominal exchange rate flexibility. The following quote from a speech of the Governor of the Bank of England (George, 1995) is not unrepresentative:

> This longer-term problem of unemployment reflects, at least in part, structural features of the European labour market, which also differ from one country to another—for example in the degree of flexibility in wages and other conditions of employment, or in the degree of non-wage, social costs of employment. It is being addressed, variously, through structural policies nationally and through measures such as those that are being explored by the European Commission and debated by the European Council. But it will not easily go away. And it could in fact become more difficult to resolve within monetary union as a result of on-going differences between member countries, for example, as a result of differences in rates of productivity growth, or unrelated differences in earnings growth, or as a result of divergent demographic trends and associated differences in dependency ratios.

The fundamental misunderstanding, reflected in the above quote, of what nominal exchange rate flexibility can deliver prompts the following proposition.

PROPOSITION 6. *Real convergence or divergence is irrelevant for monetary union.*

Asserting the contrary would mean attributing to monetary policy (under which I include exchange rate policy) power and significance well beyond what it can deliver. Does anyone really believe that the problems of Italy's Mezzogiorno would have been alleviated if Southern Italy had been given its own currency and had decided to float the southern Lira independently of the northern Lira? Or that Appalachia would have been more prosperous if it had been granted its own currency? How would real wage rigidities be alleviated by having an independent currency and a floating exchange rate? How are the competitiveness problems associated with excessive nonwage labor costs mitigated by having a floating exchange rate? Why would international differences in the severity of intergenerational distribution problems, and in the strains put on public sector budgets by graying populations and emerging "youth deficits," be any less with a floating exchange rate than under a fixed rate? There is no reason whatsoever why regions characterized by persistent differences in total factor productivity growth, or by persistent differences in real earnings growth unrelated to productivity growth differentials, cannot be locked together in a common currency area. No doubt real economic performance would be dismal in a region whose real earnings growth systematically exceeded its productivity growth, but it would be equally dismal regardless of the exchange rate regime.

With nominal inertia, monetary policy can influence the current short real interest rate; that is, it can influence the short real interest rate in the short run. With the myopia, herd-instinct and bandwagon effects that often dominate financial markets on a day-to-day basis, monetary policy may also have a transitory effect on the current long real interest rate; that is, it may be able to influence long rates in the short run (although not necessarily in a very predictable manner). Unless the economy is hysteretic, monetary policy ultimately cannot influence either the short-term real rate or the long-term real rate. *Mutatis mutandis*, the same holds for the ability of monetary and exchange rate policy to influence the real exchange rate or any other real variable.[28]

6. Restrictions on Capital Mobility

Virtually all the arguments given in Emerson et al. (1990) to the effect that the logic of market integration implies the need for a common currency are seriously flawed. Many seem to derive from fears that competitive devaluations of the nominal exchange rate can buy a country a lasting competitive advantage (a lasting real devaluation), thus distorting the competitive "level playing field" (whatever that is). If the economy has the natural rate property at least in the long run, these fears are overstated. In addition the historical evidence of the OECD countries and a wide range of developing countries and transition economies supports the view that it is not possible to gain any enduring competitive advantage by pursuing deliberately inflationary policies.

Only one aspect of market integration does indeed point in the direction suggested by the "One market, one money" school of thought. That aspect is financial market integration, and specifically the removal of fiscal and administrative obstacles to the international movement of financial capital. The key point here can be summarized in the following proposition

PROPOSITION 7. *With unrestricted international mobility of financial capital, a common currency becomes, at the very least, desirable. It may well become unavoidable.*

The arguments supporting this position are both theoretical and empirical. Managed exchange rate regimes, including fixed-but-flexible exchange rate regimes such as Bretton Woods, or target zones with hard barriers such as the original ERM, break down with probability 1 in finite time. They are not sustainable in the longer term. Floating exchange rate regimes may be feasible, but are likely to have undesirable operating characteristics: they frequently are characterized by excess short-term volatility and persistent medium-term misalignments.

Take a fixed exchange rate regime as epitome of all managed exchange rate regimes. Any fixed exchange rate regime that is not irrevocably fixed (that is, anything short of monetary union[29]) can be abandoned for one of two reasons. The authorities can choose to abandon the fixed parity, for any number of virtuous or opportunistic reasons, even in the absence of speculative attacks, or they can be forced off the fixed parity by a speculative attack that exhausts their international reserves and credit lines. While technically (that is, in a world with credible commitment) any solvent government should be able to borrow infinite amounts of foreign exchange (simply by swapping it for its own currency or debt), in reality there is a limit to the credit lines that any monetary authority can draw on. Any finite limit can be challenged by private speculators in reasonably efficient financial markets. Again, in principle a government could always raise short-term interest rates to the point where speculators decide not to short its currency. In practice, with imperfect credibility, the interest rate levels that may be required to do the trick are likely to be politically unsustainable, especially in countries like the UK where the cost of many variable rate mortgages is tied closely to the short rate of interest.

The gold standard survived as long as it did for two reasons. First, the degree of international capital mobility was less than it is today. Second, the key national authorities were not held responsible for real macroeconomic performance (output and unemployment) and could make the defense of the gold standard their overriding priority. At least since World War II, no government has been able to enjoy the luxury of focusing monetary and fiscal policy exclusively on the defense of the external value of their currency. Any commitment to a fixed parity is therefore vulnerable, and will be tested by the markets.

The completion of one component of the single market program, the elimination of (virtually) all remaining restrictions on the intra-EC mobility of financial capital, was sufficient to seal the fate of the EMS and the ERM. With all legal restrictions removed and much of the accumulated inefficiency of the previously protected private financial sectors swept away, a market mechanism was created that could shift literally hundreds of billions of dollars worth of financial claims between currencies in a matter of minutes, and at very little cost. Add to this a renewable population of unskilled and unsuccessful speculators (including those in charge of macroeconomic policy in the national ministries of finance and central banks, but also new and inexperienced players from the private sector), and all the elements for a successful attack on a fixed-but-adjustable exchange rate arrangement like the ERM were in place. This is not to say that the EMS and the ERM would have survived if effective capital controls had been in place. The literature on voluntary (or opportunistic) abandonment of the peg does not require that there be *any* international mobility of capital.

Is it possible to put the genie back in the bottle through fiscal or administrative capital controls? The scope and efficiency of the global industry ready to take on the authorities by supplying the means to avoid and evade controls is awesome. The rewards from taking on the monetary authorities are too high: given the ineffective penalties likely to be imposed and the low risk of being caught evading the controls, the odds on capital controls working effectively are virtually nil.

Proposals for imposing non-interest-bearing reserve requirements on balances used for taking open positions to attack currencies are vulnerable because they ignore key developments of the last two decades in the international financial markets. There are myriad ways now of attacking a currency: through the spot markets, through the futures markets, through the derivatives markets. "Tobin" taxes[30] on foreign exchange transactions would likewise have to be expanded in their coverage to include transactions in the option markets and in markets for all other kinds of derivatives. Note that this argument for a common currency extends beyond the EU and applies to any countries linked by unrestricted financial capital mobility, including the USA and Japan.

7. The Maastricht Convergence Criteria

A common currency may be the only logical option left after the abolition of capital controls; the question is, can we get there from here? In particular, how do the Maastricht convergence criteria enhance or impede the process?

Of the four convergence criteria (the exchange rate criterion, the interest rate criterion, the inflation criterion, and the excessive deficits criterion (which consists of a government deficit ceiling and a government debt ceiling)) only one, the exchange rate criterion, makes any sense. The interest rate and inflation criteria are largely redundant and irrelevant. Fortunately, these two are unlikely to create serious compliance problems for most European countries. The budget deficit and debt criteria are unnecessary and potentially dangerous for the EU as a whole and for individual countries. The good news is that while the exchange rate, interest rate and inflation criteria are quite specific, the deficit and debt criteria, *as specified in the Treaty of Maastricht*, are quite vague and flexible. It is important to recognize that flexibility and to take advantage of it.

Consider the four convergence criteria in turn.

The Exchange Rate

The rationale for ruling out significant parity changes (devaluations) prior to monetary union surely is to avoid the risk of "endgame" devaluations aimed either at achieving a transitional competitive advantage or at reducing the real value of public debt denominated in domestic currency.

As long as monetary autonomy is expected to exist in the future, maintaining a reputation for being tough on inflation is valuable to the monetary authority. The cost of losing that reputation militates against the temptation to gain a competitive advantage (or amortize debt) through devaluation. Once monetary union is a fact, national reputations for monetary restraint are worth nothing (see Froot and Rogoff, 1991; Bayoumi, 1995). The temptation to get in one last, big devaluation before the ECB throws the key away, may be hard to resist. The exchange rate criterion rules this out and therefore makes sense from the point of view of avoiding zero-sum (at best) "endgame" devaluations in pursuit of national competitive advantage. "Endgame" devaluations as a means of amortizing excessive public debt are of course also ruled out by the exchange rate criterion.

Interest Rates

Long-term nominal rates of interest on government debt are required to converge to a level close to that achieved by the three countries with the lowest rate of inflation. In the absence of (1) differences in effective marginal tax rates on interest income, and (2) differences in default risk, there will be complete interest rate equalization immediately following currency unification. The only way to make sense of the interest rate convergence criterion, which imposes limits on interest rate spreads *prior to* monetary union, is that it is either a device for imposing interest income tax harmonization or yet another stratagem for keeping out of the monetary union governments whose debt is subject to a significantly higher default risk premium than the debt of the other member governments. Barring differential default risk and differential interest taxation, the criterion is redundant: monetary union ensures interest rate equalization—the cart is put before the horse. The interest rate criterion therefore only makes sense as a fiscal criterion in disguise.

Inflation

Prior to being allowed to join, a prospective entrant's inflation rate must be close to the inflation rates achieved by the three countries with the lowest rates of inflation. It is clear that monetary union is a means for achieving inflation convergence. Inflation rates for traded goods should converge quite quickly, while nontraded good prices and costs also would ultimately rise at a common rate (corrected for the familiar inter-member differences in the productivity growth differential between the traded and nontraded good sectors). Why then impose inflation convergence prior to monetary union as a criterion for EMU membership?

If there is a reasonable answer, it must involve an empirical judgment about the inheritability of inflation persistence (or inflation inertia) following monetary union. The issue is a fascinating and important one, and one on which there is little or no empirical evidence. Clearly, if there is no inflation persistence, the prior inflation convergence criterion makes no sense. However, in order to reach the conclusion that prior inflation convergence is desirable, it does not suffice to note that potential EMU members have historically been characterized by inflation inertia. Assume that Italy has inflation inertia and that the current core inflation rate (in lira) is x percent per annum. As long as contracts are denominated in lira, this core inflation rate will respond only sluggishly to changes in economic conditions (that is the meaning of inflation inertia). It is by no means clear, however, what will happen to Italian inflation persistence once contracts are denominated in the new currency (the euro). Will Italian euro inflation inertia simply inherit Italian lira inflation inertia, or will it instead evolve according to a different process (say the average prior core inflation of the other EMU members?). If there is full "inheritance" of national inflation inertia, conver-gence of core inflation rates prior to EMU is desirable to avoid important changes in relative prices and costs building up under EMU before national core inflation rates have converged. It probably makes sense to be cautious, but this is surely a judgment that can be left to the individual member countries and does not need to be written in stone and enforced centrally.

Public Debt and Deficit Ceilings

The two numerical fiscal criteria of the Maastricht Treaty make no sense and should be jettisoned. Some of the reasons are as follows.

First, the two figures are arbitrary and their origins accidental. Sixty percent hap-pened to be the average debt–GDP ratio in the EC at the time the Maastricht Treaty was being cobbled together. A 3% deficit–GDP ratio is consistent with the mainte-nance of a 60% debt–GDP ratio[31] when the growth rate of nominal GDP is 5% per annum (say 3% real growth and 2% inflation). No normative significance attaches to a 5% growth rate of nominal GDP. Even if we think a 5% annual growth rate of nominal GDP to be a desirable EU-wide target, why a debt ratio of 60% with a deficit ratio of 3%, rather than a debt ratio of 80% with a deficit ratio of 4% (which would also be consistent with each other), or perhaps a debt ratio of 20% with a deficit ratio of 1%?

Other things being equal, I would like most EU governments to have less debt and a smaller deficit. In the best of all possible worlds, the state would not be a debtor at all but a creditor, who finances all socially necessary expenditures out of the interest income it receives from its financial assets! That would do away with the disincentive effects of distortionary taxation, and in addition it would boost the national saving rate. The key question, however, is not whether we could have done better and should do better in the future. It is rather whether *now* (that is before the beginning of 1998 when the list of countries satisfying all the criteria for EMU is drawn up) is the time to engage in fiscal heroics to bring down the average EU debt–GDP ratio significantly. The answer for most would-be EMU members is a clear "no."

Second, higher debt and deficit levels are feasible (sometimes even sustainable), and may, under specific circumstances, be desirable. Immediately following World War II the UK debt–GDP ratio was over 250%. The highest ever figure was recorded following the Napoleonic wars. What is an appropriate debt and deficit level depends on economic structure (e.g., private saving rates and the level of private financial wealth), on historical circumstances (the aftermath of war or deep recession), and on the role accorded to public debt and deficits in intergenerational redistribution through the budget. To have a "one-size fits all" upper limit to public debt and deficit ratios is an economic nonsense.

Third, the one instance of a monetary union within the EU, that between Belgium and Luxembourg (since 1922), combines the countries with the highest and the lowest debt–GDP ratios in the EU. Admittedly, it is difficult to extrapolate from a currency union between a small country and a dwarf state to a union between a number of medium-sized and small countries. Taking a longer historical perspective, it is interesting that the Dutch Republic (1572–1795) combined monetary union with fiscal autonomy for the seven provinces, a very limited federal budget and the complete absence of a proper federal tax system (Dormans, 1991).

Fourth, while the deficit is defined in terms of *net* borrowing, the debt concept is defined in terms of *gross* debt. Financial window-dressing (liquidating financial assets to pay off financial debt) is invited. The privatization of public sector enterprises should be undertaken for efficiency reasons, not for revenue reasons (at any rate in advanced industrial countries), and certainly not for financial window dressing.[32] More generally, focusing on conventionally measured public debt aggregates diverts attention from the many "off-balance sheet" implicit contingent liabilities of the government (future social security benefits, state pensions, loan guarantees, export credit guarantees), and from other future spending commitments that may be a more serious threat to fiscal probity and solvency. Off-balance sheet public sector assets (natural resource or mineral rights) are likewise ignored.

Fifth, the criterion fails to allow properly for the distinction between public sector consumption spending and public sector capital formation, although the Treaty actually contains a rather vague reference to the effect that "The report of the Commission shall also take into account whether the government deficit exceeds government investment expenditure . . ." (Article 104c, 3).

Sixth, with the average debt–GDP ratio in the EU now above 70% of GDP and the average deficit at 4.5% of GDP, a serious attempt to meet the criteria by 1998 would require a major fiscal contraction. This should be contemplated only if an offsetting EU-wide relaxation of monetary policy were contemplated, permitting the overall level of aggregate demand to be maintained while the change in the policy mix favors a change in the composition of demand towards investment and away from private and/or public consumption. The assertion that fiscal contractions, far from depressing aggregate demand, will boost it ("crowding in" rather than "crowding out"), either through "confidence" effects or through the favourable *announcement effects* of future fiscal contractions on the bond market and the stock market, is central bankers' pie-in-the-sky, except possibly in countries where the fiscal situation is so out of hand that there is a serious threat of state default. Only Belgium, Italy, and Greece are potentially in that category.

Anticyclical deficits are desirable both for Keynesian demand-management reasons and because it is efficient to finance temporary increases in public spending in part by borrowing, thus minimizing the need for large fluctuations over time in distortionary marginal tax rates. Note that it is not correct to say that a strict interpretation of the deficit criterion permanently immobilizes the automatic fiscal stabilizers. After all, deficits can fluctuate around an average value of zero or minus 2% of GDP as easily as around an average value of 3% or 4% of GDP. A strict interpretation of the deficit criterion as an upper limit would, however, immobilize the automatic stabilizers for a number of years, with the EU average deficit already at 4.5% of GDP.

The debt and deficit criteria do not constitute a mechanism for coordinating fiscal policy in the EU as a whole, let alone for achieving the right EU-wide monetary–fiscal policy mix. This is something on which the Treaty is silent. A mechanism or procedure for ensuring the right overall monetary–fiscal policy mix is indeed hardly compatible with the notion of central bank independence enshrined in the Treaty—and constitutes a powerful argument against central bank independence.

Fortunately, the debt and deficit criteria were applied quite sensibly and flexibly in the one real test case we have had thus far, that of Ireland. In 1995 Ireland's government deficit was 2.7% of GDP and its government debt 85.9% of GDP. The corresponding figures for 1991 were 2.2% and 96.7% respectively. The Irish were judged to have passed the test. That is the kind of flexibility the EMU program can survive with.[33]

What then is the purpose of the excessive deficits criterion? One somewhat cynical interpretation is that it was included to provide Germany with an alibi if it decides it does not wish to give up the D-mark when the time comes. A more charitable interpretation is that it is intended to strengthen the hand of the European Central Bank *vis-à-vis* the national ministries of finance, and the hand of the "fiscally responsible" countries and of Brussels *vis-à-vis* "fiscally irresponsible" countries like Italy, Greece, and Belgium. The fiscal criterion aims to strengthen the effectiveness of the "no bail-out" (directly by other ministries of finance or by Brussels and indirectly by monetization through the ECB) clause by making it less likely that a debt default contingency would ever arise or that any country could ever blackmail the rest of the EU into servicing part of its debt.

I view the fear that a fiscally irresponsible EMU member (say, Italy, for the sake of argument) could blackmail the rest of the Union (or the ECB) into effectively taking over some of its debt as overblown and unrealistic. After all, what could Italy threaten the rest of the EMU countries with? A rescheduling of or default on some or all of its debt? That is first and foremost a distributional issue between on the one hand the holders of the Italian public debt and on the other hand the Italian taxpayers and the beneficiaries from Italian public spending.

Are there likely to be systemic externalities (e.g., an EMU-wide financial crisis) from an Italian rescheduling or default? Investors in Italian securities have earned risk premia for quite a while now. Careful portfolio managers will have realized that risk premia tend to be paid because there is risk and will have built up reserves to allow for unfavorable contingencies, including default. If there has not been adequate provisioning against default risk, major Italian bond holders, including banks, could be faced with default risk. The solution to that problem is *banking supervision and regulation*, including upper limits on the exposure of any financial institution to sovereign default risk, not the imposition of debt and deficit ceilings on national governments.

If worst came to worst and a commercial bank rescue operation by the European Central Bank were required, the integrity of the banking system (and especially of the payment mechanism) could be salvaged without bailing out the Italian state. After all, the ECB should be a more effective debt collector than the private holders of Italian government debt. Note also that the Italian state would be in a much weaker bargaining position *vis-à-vis* the European Central Bank than it is today *vis-à-vis* its own central bank. The risk of sovereign default, or of central bank monetization of the public debt, would be less with EMU than without it.

It is essential that the ECB be willing and able to act as *lender of last resort* should the need arise. The Treaty is, unfortunately, almost silent on this key role of any central bank. Moral hazard always rears its head when the lender of last resort function is taken seriously, but it can be minimized by the central bank adopting the following, familiar operating rules. (1) For solvent financial institutions faced with a liquidity crisis, lend freely but at a penal rate. (2) For insolvent financial institutions whose sudden demise would cause serious systemic externalities, first ensure that all equity claims are extinguished as soon as any public money goes in and, second, fire the entire top management of the institutions without any golden handshakes.

It follows from this analysis that Waigel's proposal for a "stability pact" (a tighter version of the Maastricht fiscal criteria) for EMU members following monetary union has even less to recommend it than the original Maastricht fiscal criteria. Compulsory interest-free deposits with the ECB, and eventual forfeit of these deposits if the Waigel criteria continued to be flouted, would add injury to insult.[34]

As long as the fiscal criteria are applied sensibly and flexibly, *and as long as the overall fiscal–monetary policy mix in the EU as a whole is adjusted appropriately*, no serious damage need result from the pursuit of the Maastricht fiscal norms as a medium-term goal. The automatic fiscal stabilizers can continue to perform their normal cyclical stabilizing functions at the national level. Each national government can aim to reduce its claim on its national savings in ways and at a rate that respect differences in initial conditions, economic structures, and external environments, supported by an appropriate EU-wide monetary policy aimed at maintaining aggregate demand in the face of whatever fiscal retrenchment is deemed necessary— not because of the arbitrary Maastricht criteria, but for sound national economic reasons.

8. Conclusions

Most economic arguments for or against monetary union are misconceived and overstated. All the heated talk about the international monetary and exchange rate system should not blind us to the fact that it is merely monetary policy, or rather the international dimension of monetary policy, that is at stake. Monetary policy is not unimportant from the point of view of short-run, cyclical stabilization, but neither is it the stuff of which the wealth of nations is made.

The narrowly economic case for monetary union seems quite finely balanced. As stated in the Introduction, it is unlikely that the issue will be decided on anything else than general political grounds. If the political tide towards greater European integration is halted or reversed, especially in Germany, monetary union will go by the board.

If European integration gets a second wind, much as it did in 1986 with the signing of the Single European Act, there will be a single European currency for most of the current EU members by the beginning of the next century.

Such subordination of economics to politics may be sobering for economists. It fits in well with the long history of the quest for exchange rate stability in Europe. It is also very much in the spirit of the interpretation of the causes of the breakdown of the ERM in 1992/93 proposed in Buiter et al. (1997). In that interpretation, the principal cause of the collapse of the system in 1992/93 is a quintessentially political event: the breakdown of cooperative behavior among the Periphery countries when faced with an exogenous shock—the boost to aggregate demand in the Center—joined to the continued (and long-standing) unwillingness of the Center to compromise on its domestic macroeconomic objectives. It now seems certain that what began with politics will also end with politics. *Plus ça change . . .*

References

Bayoumi, Tamim, "Who Needs Bands? Exchange Rate Policy Before EMU," CEPR discussion paper 1188, 1995.
Bayoumi, Tamim and Barry Eichengreen, "Shocking Aspects of European Monetary Unification," in Francisco Torres and Francesco Giavazzi (eds.), *Adjustment and Growth in the European Monetary Union*, Cambridge: Cambridge University Press, 1993.
Bayoumi, Tamim and Paul Masson, "Fiscal Flows in the United States and Canada: Lessons for Monetary Union in Europe," CEPR discussion paper 1057, 1994.
Bayoumi, Tamim and Alun Thomas, "Relative Prices and Economic Adjustment in the United States and the European Union: a Real Story about EMU," *IMF Staff Papers* 42(1) (1995):108–33.
Bini-Smaghi, Lorenzo, Tommaso Padoa-Schioppa and Francesco Papadia, "The Transition to EMU in the Maastricht Treaty," *Essays in International Finance* 194, International Finance Section, Princeton University, 1994.
Blanchard, Olivier and L. Katz, "Regional Evolutions," *Brookings Papers on Economic Activity* 1, 1992.
Buiter, Willem H., "Measurement of the Public Sector Deficit and its Implications for Policy Evaluation and Design", *IMF Staff Papers* 30 (1983a):306–49.
———,"The Theory of Optimum Deficits and Debt," in Federal Reserve Bank of Boston, *The Economics of Large Government Deficits*, Conference Series 27 (1983b):4–69.
———,"International Monetary Policy to Promote Economic Recovery," in C. van Ewijk and J. J. Klant (eds.), *Monetary Conditions for Economic Recovery*, Amsterdam: Martinus Nijhoff, 1985. Reprinted in Willem H. Buiter, *International Macroeconomics*, Oxford: Oxford University Press, 1990.
———,"Politique Macroéconomique dans la Période de Transition vers l'Union Monétaire," *Revue d'Economie Politique* 105 (1995):807–46.
Buiter, Willem H. and Marcus H. Miller, "Cost and Benefits of an Anti-Inflationary Policy: Questions and Issues," in V. E. Argy and J. W. Nevile (eds.), *Inflation and Unemployment: Theory, Experience and Policy-Making*, London: George Allen & Unwin, 1985:11–38.
Buiter, Willem H., Giancarlo Corsetti, and Paolo Pesenti, "A Center–Periphery Model of Monetary Coordination and Exchange Rate Crises," NBER working paper 5140, 1995.
———, *Financial Markets and European Monetary Cooperation: The Lessons of the 92–93 ERM Crisis,* Cambridge: Cambridge University Press, 1997, forthcoming.
Canzoneri, Matthew B. and Carol Ann Rogers, "Is the European Community an Optimal Currency Area? Optimal Taxation Versus the Cost of Multiple Currencies," *American Economic Review* 80 (1990):419–33.

Chamley, Christopher, "Optimal Taxation of Capital Income in General Equilibrium with Infinite Lives," *Econometrica* 54 (1986):607–22.

Corsetti, Giancarlo, "Public Finance in Stochastic Models of Endogenous Growth," manuscript, Yale University, 1992.

Corsetti, Giancarlo and Paolo Pesenti, "Growth with Risky Assets: Money and Budgetary Policy in a Portfolio Model of Endogenous Growth," manuscript, Princeton University, 1992.

Courchene, T. J., "Reflections on the Canadian Federalism: Are there Implications for the European Economic and Monetary Union?" *European Economy*, Special Issue 4 (1993):23–166.

De Grauwe, Paul and Wim Vanhaverbeke, "Is Europe an Optimum Currency Area? Evidence from Regional Data," CEPR discussion paper 555, 1991.

De La Dehesa, G. and P. Krugman, "Monetary Union, Regional Cohesion and Regional Shocks," in G. De La Dehesa, A. Giovannini, M. Guitian and R. Portes (eds.), *The Monetary Future of Europe*, London, Centre for Economic Policy Research.

Dormans, E. H. M., *Het Tekort*, Amsterdam, NEHA-Series III, 1991.

Dowd, Kevin and David Greenaway, "Currency Competition, Network Externalities and Switching Costs: Towards an Alternative View of Optimum Currency Areas," *Economic Journal* 102 (1993):1180–9.

Drazen, A., "A General Measure of Inflation Tax Revenues," *Economics Letters* 17 (1985):327–30.

Eichengreen, Barry, "One money for Europe? Lessons from the US Currency Union," *Economic Policy* 10 (1990a):117–87.

———,"Is Europe an Optimum Currency Area?" CEPR discussion paper 478, 1990b.

Eichengreen, Barry and Charles Wyplosz, "The Unstable EMS," *Brookings Papers on Economic Activity* 1 (1993):51–124.

Emerson, Michael, Daniel Gros, Alexander Italianer, Jean Pisani-Ferry, and Horst Reichenbach, "One Market, One Money," *European Economy* 44 (1990). Also published by Oxford University Press, 1992.

van Ewijk, Casper, "The Distribution of Seigniorage: A Note on Klein and Neumann," *Weltwirtschaftliches Archiv* 128 (1992):346–51.

Financial Times, "Foreign D-Mark Holdings Swell," Wednesday 21 June 1995, p. 2.

Froot, Kenneth A. and Kenneth Rogoff, "The EMS, the EMU, and the Transition to a Common Currency," in Olivier J. Blanchard and Stanley Fischer (eds.), *NBER Macroeconomics Annual 1991*, Cambridge, MA: MIT Press, 1991:269–317.

George, E. A. J., "The Economics of EMU," Churchill Memorial Lecture, 21 February, 1995.

Goodhart, C. A. E. and S. Smith, "Stabilization," *European Economy*, Special Issue 5 (1993):417–56.

Grilli, Vittorio, "Exchange Rates and Seigniorage," *European Economic Review* 33 (1989a):580–7.

———, "Seigniorage in Europe," in Marcello de Cecco and Alberto Giovannini (eds.), *A European Central Bank? Perspectives on Monetary Unification After Ten Years of the EMS*, Cambridge: Cambridge University Press, 1989b:53–79.

Ingram, James, "The Currency Area Problem," in R. A. Mundell and A. K. Svoboda (eds.), *Monetary Problems of the International Economy*, Chicago, IL: Chicago University Press, 1969.

———, "The Case for European Monetary Integration," *Princeton Essays in International Finance* 98, 1973.

Kenen, Peter, "The Theory of Optimum Currency Areas: An Eclectic View," in R. A. Mundell and A. K. Swoboda (eds.), *Monetary Problems of the International Economy*, Chicago, IL: University of Chicago Press. 1969.

Klein, Martin and Manfred J. M. Neumann, "Seigniorage: What Is It and Who Gets It?" *Weltwirtschaftliches Archiv* 126 (1990):205–21.

Klein, Martin and Manfred J. M. Neumann, "The Distribution of Seigniorage," *Weltwirtschaftliches Archiv* 128 (1992):52–356.

Krugman, P., "Second Thoughts on EMU," *Japan and the World Economy* 4 (1992).

———, "Lessons of Massachusetts for EMU," in F. Torres and F. Giavazzi (eds.), *Adjustment and Growth in the European Monetary Union*, Cambridge: Cambridge University Press (1993).

Leeftinck, Bertholt, "The Desirability of Currency Unification: Theory and Some Evidence," *Tinbergen Institute Research Series* 92, 1994.

Lucas, Robert E., Jr, "Supply-Side Economics: An Analytical Review," *Oxford Economic Papers* 42 (1990):293–316.

McKinnon, Ronald I., "Optimum Currency Areas," *American Economic Review* 53 (1963):717–25.

Mantel, Sophie, "The Prospects for Labour Mobility under EMU," *Economie et Statistiques*, Special Issue (1994):137–47.

Masson, Paul R. and Mark P. Taylor, "Common Currency Areas and Currency Unions: An Analysis of the Issues," CEPR discussion paper 44, 1992.

Melitz, Jacques, "A Suggested Reformulation of the Theory of Optimal Currency Areas," CEPR discussion paper 590, 1991.

Muet, Pierre-Alain, "Croissance, Emploi et Chomage dans les Années Quatre-Vingt," *Revue de l'OFCE* 35 (1995).

———, "Ajustements Macroéconomiques et Coordination en Union Monetaire," paper presented at the Journées AFSE 1995, Nantes, 8–9 June 1995.

Mundell, Robert A., "A Theory of Optimum Currency Areas," *American Economic Review* 51 (1961):657–75.

Nikolakaki, Maria, "Is Europe an Optimum Currency Area? A Reconsideration of the Evidence," manuscript, London School of Economics, 1996.

Poole, William, "Optimal Choice of Monetary Policy Instruments in a Simple Stochastic Macro Model," *Quarterly Journal of Economics* 84 (1970):197–216.

Roubini, Nouriel and Gian Maria Milesi-Ferretti, "Optimal Taxation of Human and Physical Capital in Endogenous Growth Models," NBER working paper 4882, 1994.

Rovelli, Riccardo, "Reserve Requirements, Seigniorage and the Financing of the Government in an Economic and Monetary Union," *European Economy* 1 (1994):11–55.

Sala i Martin, Xavier and Jeffrey Sachs, "Fiscal Federalism and Optimum Currency Areas: Evidence for Europe from the United States," CEPR discussion paper 632, 1992.

Tobin, James, "A Proposal for International Monetary Reform," ch. 20 in J. Tobin, *Essays in Economic Theory and Policy*, Cambridge, MA: MIT Press, 1982.

Van Rompuy, Paul, Filip Abraham, and Dirk Heremans, "Economic Federalism and the EMU," *European Economy* Special Issue 1 (1991):109–35.

Von Hagen, Jurgen, "Fiscal Arrangements in a Monetary Union: Some Evidence from the US," in Don Fair and Christian de Boissieu (eds.), *Fiscal Policy, Taxes and the Financial System in an Increasingly Integrated Europe*, Deventer: Kluwer, 1992.

Von Hagen, Jurgen and George W. Hammond, "Regional Insurance Against Asymmetric Shocks: An Empirical Study for the European Community," CEPR discussion paper 1170, 1995.

Zhu, Xiadong, "Optimal Fiscal Policy in a Stochastic Growth Model," *Journal of Economic Theory* 58 (1992):250–89.

Notes

1. The nonconvexity intrinsic in this transaction technology means that the value of the technological network externalities is not captured by the bid–ask spread in the foreign exchange markets, even if the latter are competitive and efficient.

2. An updated analogy would emphasize the social gains of adopting common or compatible software systems.

3. See Dowd and Greenaway (1993). Note that a related kind of switching cost is also present whenever a multiple currency system is in effect. Boundedly rational agents will incur temporary switching costs every time they have to switch between *numéraires*.

4. Lest I be accused of having feet planted firmly in the air, I am aware that considerations of national pride and international envy are likely to rule out any efficient solution to the problem of the name of the common European currency proposed here.

5. In this section "government" refers to the consolidated general government and central bank.

6. The same applies to other proposed measures of the resource transfer between the general government and the central bank such as the *fiscal seigniorage* favored by Klein and Neumann (1990, 1992), and by Rovelli (1994). Klein and Neumann define fiscal seigniorage as central bank profits transferred to the general government, plus the net increase in central bank credits to the government, minus interest payments of the government to the central bank. As pointed out by van Ewijk (1992), there are questions about the appropriateness of this definition even as a measure of the contribution the general government gets from the central bank's monopoly over base money. Drazen (1985) consolidates the general government sector with the central bank and consequently has a measure of seigniorage which corresponds to the one used here (Buiter, 1983a, 1983b).

7. We are talking potentially serious money. A recent Bundesbank study, reported in the *Financial Times* (1995), estimated that some 30–40% of the total currency circulating outside the banking system (between DM65bn and DM90bn) was probably abroad. The corresponding figure for the US dollar was estimated to be between 60% and 70%.

8. They are to be distinguished from the non-neutralities that would be present even in a world without nominal rigidities and that reflect the effects of anticipated inflation on consumption demand and portfolio allocation. "Superneutrality" of money, that is, invariance of real equilibrium allocations in classical competitive equilibrium models with complete markets, under alternative fully anticipated rates of growth of the nominal money stock and associated rates of inflation, is unlikely to be of much practical interest. The Mundell–Tobin effect is probably the best-known channel through which higher anticipated inflation, by affecting the portfolio choice between money and real capital, influences real equilibrium allocations. It will be ignored in what follows.

9. Economic theory has been unable so far to provide a convincing rationalization of this particular form of incomplete contracting. Indeed, we do not have good theories as to why wage and price contracts tend to use money (the medium of exchange and means of payment) as the *numéraire* (unit of account), rather than some other bundle of goods.

10. For simplicity I consider a deterministic steady state.

11. Including the rate of inflation, which is of course a real variable.

12. As we shall see below, the transitory difference made by nominal exchange rate flexibility to the real adjustment path of the economy is potentially desirable in the case of shocks to goods market demand, but potentially undesirable in the case of monetary shocks.

13. The classic references are Mundell (1961); McKinnon (1963); Kenen (1969); Ingram (1969, 1973); Ishiyama (1975). Among recent contributions, see Melitz (1991); De Grauwe and Vanhaverbeke (1991); Masson and Taylor (1992); Krugman (1992); Krugman (1992, 1993); Dehesa and Krugman (1993); Eichengreen (1990a, 1990b); Bayoumi and Eichengreen (1993); Bayoumi and Thomas (1995); Bini-Smaghi and Vori (1993); Eichengreen and Wyplosz (1993); Leeftinck (1994), Bayoumi (1995); von Hagen and Hammond (1995) and Muet (1995).

14. In the model under consideration, labor services (and leisure) are of course nontraded goods, so variations in the nominal exchange rate still work by influencing the relative price of traded and nontraded goods. The only other transmission channel would be the asset revaluation effects of nominal exchange rate changes, including real balance effects.

15. Semi-small because it faces a downward-sloping demand curve for exportables, while it treats the world nominal rate of interest and the foreign price level as parametric.

16. Strictly speaking, λ^l is a shock to the excess demand for money. For our purposes, it does not matter whether it is a shock to money demand or to money supply.

17. Alternatively, the objective could be to stabilize output around its *ex ante* full information natural level, 0. The optimal policy response to LM shocks and IS shocks is unaffected by this. The optimal policy response to supply shocks obviously would be.

18. It is well-known, that the optimal policy in the face of a range of shocks is a "combination policy" expressing the exchange rate as a function of the nominal interest rate. Pegging the interest rate and pegging the quantity of money are two special cases of this. I focus on these two special cases because they seem the most operationally relevant: the authorities are not, in my view, likely to be able to identify the (time-varying) variance–covariance structure of the fundamental shocks to the economy.

19. The complete two-country model adds the following equations to equations (10) to (16):

$$m_t^* - q_t^* = k^*\left(p_t^* + y_t^* - q_t^*\right) - \gamma^* i_t^* + \lambda_t^{*1} \qquad y_t^* = -v^* r_t^* + \delta z_t + \lambda_t^{*d}$$

$$r_t^* \equiv i_t^* - E_t q_{t+1}^* + q_t^* \qquad q_t^* \equiv (1 - \beta^*) p_t^* + \beta^*\left(p_t + s_t\right)$$

$$y_t^* = \alpha^*\left(p_t^* - E_{t-1} p_t^*\right) + \lambda_t^{*s} \qquad m_t^* = \theta^* d_t^* + (1 - \theta^*) p_t.$$

20. Strictly speaking the model is hysteretic in the values of the nominal variables rather than indeterminate. In period t, $E_{t-1} p_t$ and all other expectations, formed in period $t - 1$, of future nominal prices are predetermined and will provide a nominal anchor for the model in period t.

21. Having a nonzero stock of domestic-currency denominated interest-bearing public debt outstanding would also suffice to eliminate the price level indeterminacy problem. Neutrality (or homogeneity of degree zero of the real equilibrium) applies only to all outside nominal asset stocks, not to money alone.

22. Note that this is a different point from the usual Lucas critique, which is discussed below.

23. IS shocks can have a long-run effect on the real exchange rate but not on real output. LM shocks affect neither real output nor the real exchange rate in the long run.

24. See, e.g., Eichengreen (1990a, 1990b), Muet (1991, 1995), Blanchard and Katz (1992), and Mantel (1994).

25. Indeed, the loss of the national monetary instrument could, in principle, be compensated for through the more active use of domestic fiscal instruments (including the automatic stabilizers), without any need for international transfers. The Maastricht fiscal criteria make this difficult, with both the average EU general government deficit (about 4.5% of GDP in 1995) and the average general government gross debt (over 70% of GDP in 1995) above their ceilings of 3% of GDP and 60% of GDP respectively.

26. In the empirical analysis of Sala i Martin and Sachs (1992), there is a large area of ambiguity between insurance against certain kinds of transitory shocks (which is all that nominal exchange rate flexibility can provide) and (potentially) permanent redistribution through the Federal Budget. See also Eichengreen (1990a), Van Rompuy, Abraham and Heremans (1991), von Hagen (1992), Courchene (1993), Goodhart and Smith (1993), Bayoumi and Masson (1994), and Muet (1995).

27. It may well be that much greater permanent or structural international and interregional redistribution will be required within the EU in order to render the system politically viable. That, however, is a quite separate matter from the issue of what needs to be done in order to make up for the loss of the national exchange rate instrument.

28. Other than the nominal interest rate, which is, despite its name, another real variable.

29. It is true that even monetary union is not irreversible. The Maastricht Treaty does not, however, have any provisions for a country leaving EMU after joining it. Indeed, neither the Rome Treaty nor the Maastricht Treaty have provisions for member states leaving any of the European institutions to which they have acceded.

30. See, e.g., Tobin (1982).

31. Ignoring the gross–net confusion referred to below.

32. While privatization receipts are not supposed to be netted against the Maastricht measure of the deficit, recent French legerdemain with the pension liabilities of a major French public enterprise shows that those determined to cheat can and will do so.

33. In 1996, only Luxembourg, Denmark, and Ireland were judged to have passed the "excessive deficits" test.

34. With thanks to Marcus Miller.

Review of International Economics, Special Supplement, 36–54, 1997

Wage Rigidity, Monetary Integration and Fiscal Stabilization in Europe

*Svend E. Hougaard Jensen**

Abstract

Flexibility in fiscal policy is a necessary ingredient in a policy package for EMU. Even with strong endogenous shock absorbers, such as real wage flexibility, fiscal policy can speed up the stabilization process in response to demand shocks. If real wages are rigid, as they typically are in Europe, fiscal policy cannot remove the adverse effects of asymmetric supply shocks, but it can successfully limit the divergence between member states. Monetary flexibility, a possible option in the run-up to EMU, cannot completely make up for the stabilization function of fiscal policy.

1. Introduction

As monetary integration progresses during the transition to a single currency in Europe, monetary policy will increasingly be directed at meeting the exchange rate convergence criterion. Likewise, in order to qualify for entrance into the final stage of EMU, fiscal policy will have to be conducted so as to respect certain upper limits on government debt and deficits. Since only very few EU economies currently respect these fiscal criteria, most countries will have to change their fiscal policies considerably in order to qualify for full membership. In effect, not only does monetary autonomy come to an end, also fiscal policy may be prevented from fulfilling its stabilization function.[1]

This assignment may, for several reasons, be inappropriate. First of all, adherence to a rigid EMU system of monetary and fiscal policy may force Europe into deflation and recession. Given the current high levels of unemployment in Europe, a design-risk of fiscal overkill should therefore not be overlooked. Furthermore, it is not clear that alternative shock absorbers, such as wage–price flexibility and factor mobility, can easily be made to function. Indeed, there is widespread evidence of a high degree of real wage rigidity in Europe, and, at present, labor mobility between the EU countries is rather limited. In view of such labor market rigidities, the role of fiscal policy as an exogenous shock absorber seems to become all the more important.

This paper explores in further detail the interaction between macroeconomic policy design and wage formation in an integrated European economy. Specifically, the paper discusses the design of fiscal (and monetary) policy in, respectively, a world characterized by a high degree of wage-price flexibility, where the endogenous shock absorbers are strong, and a world where output supply is constrained by a high degree of real wage rigidity. An important question is whether, and to what extent, flexibility in policy design can make up for rigidities in labor market structures, and vice versa. To illustrate a rigid policy framework I take a full-fledged monetary union with restrictions on

*Jensen: University of Copenhagen and Economic Policy Research Unit, Studiestraede 6, DK-1455 Copenhagen K, Denmark. Tel: (+45) 35323003; Fax: (+45) 35323000; Email: okoseh@pc.ibt.dk. Without implication, I thank Michael Hutchison, Neil Rankin, Bas van Aarle, two anonymous referees, and conference participants at Nantes and Innsbruck for helpful comments and discussion. Ninette Pilegaard Hansen has provided valuable research assistance. Financial support from the Danish National Research Foundation is gratefully acknowledged.

fiscal policy, and a more flexible policy regime is represented by the extended target zone proposal of real exchange rate and nominal income targeting.[2]

The device used here to address these issues is a standard two-country open-economy model with perfect international capital mobility. Assuming that the nonvertical long-run Phillips curve is empirically irrelevant to the EU, any monetary non-neutralities are short-run only. However, while both the costs and benefits of monetary union are strictly temporary, this does not mean that they can or should be neglected. On the contrary, since "temporary" may cover any time interval from one second to 15 million years, anything less than the steady state, there will in general be a case for rules designed so as to speed up the stabilization process, even if real wages are fairly flexible. Moreover, if hysteresis is present in Europe, as many believe it is, the need for flexible policies would be even stronger.

As monetary policy becomes rigid, flexibility can be achieved only through fiscal policy. I assume that fiscal policy takes the form of simple feedback rules, which may be justified because of their intelligibility and feasibility. However, they do not result from explicit optimization carried out by forward-looking authorities, and they do not explicitly respect the government's intertemporal budget constraint. Thereby I implicitly rule out potential concerns about the long-term sustainability of fiscal policies. This may be a serious drawback since one of the reasons for fiscal policy attention in the Maastricht Treaty is a perception that fiscal policies in the past have led to excessive debt accumulation. While this limitation of the model restricts the policy implications drawn from the analysis, it would still be of interest to see what short-term stabilization can achieve in response to asymmetric shocks under the implicit assumption that fiscal policy will be used sensibly.

I focus on two shocks that may seriously threaten the sustainability of a monetary union, namely an asymmetric demand shock (to the current account balances, for example) and an asymmetric supply shock, taking the form of huge wage demands (like "les évènements de Mai 1968"). Although the adjustment costs caused by asymmetric shocks may be of minor significance and diminishing *after* monetary union has been formed (European Commission, 1990), there is also evidence suggesting that country-specific shocks will remain important (Bayoumi and Eichengreen, 1993; Bayoumi and Prassad, 1995; Christodoulakis et al., 1995). What macroeconomic policy can do in response to such shocks is assessed in terms of the extent to which it contributes to (a) speed up the stabilization process for Europe as a whole; (b) alleviate the effects in the country in which the shock originates; and (c) insulate the partner's economy.

The rest of the paper is structured as follows. Section 2 outlines the details of the economic structure. Section 3 and section 4 then provide an evaluation of how, respectively, a monetary union and an extended target zone proposal perform in the face of shocks.

2. The Model

I use a dynamic two-country macro model with rational expectations.[3] The model is linear and cast in continuous time. Except for the nominal interest rate, all variables are in natural logarithms and expressed as deviations from their means. Dots denote time derivatives and subscripts refer to the particular country. The equations of the model are as follows:

$$y_i = -\beta(w_i - p_i); \quad i = 1,2, \tag{1}$$

$$w_i = \mu p_i^c + u_i; \quad i = 1,2, \tag{2}$$

$$p_i^c = p_i \pm \gamma c; \quad i = 1,2, \tag{3}$$

$$y_i = \pm \delta_1 c - \delta_2 r_i + \delta_3 y_j - \delta_4 s_i + v_i; \quad i, j = 1,2; \quad i \neq j, \tag{4}$$

$$m_i = p_i + y_i - \lambda i_i; \quad i = 1,2, \tag{5}$$

$$\dot{p}_i = vy_i + \pi_i; \quad i = 1,2, \tag{6}$$

$$\dot{e} = i_1 - i_2, \tag{7}$$

$$r_i \equiv i_i - \dot{p}_i; \quad i = 1,2, \tag{8}$$

$$c \equiv e + p_2 - p_1, \tag{9}$$

where y is real output, w the nominal wage, p the domestic output price level, p^c the consumer price index, π the expected ("core") rate of inflation, i and r the nominal and real interest rates, respectively, e and c the nominal and real exchange rates, respectively, s an index of taxation (in excess of a "neutral" budget), m the nominal money stock, u an exogenous cost-push factor, and v an exogenous shock to the demand for goods and services.[4]

Equation is the AS schedule, resulting from profit-maximizing behaviour of competitive firms. Equation (2) links the nominal wage to the consumer price index through the indexation parameter μ. The consumer price (or cost-of-living) index is given in (3) as a geometric average of domestic costs and import prices; that is, Cobb–Douglas preferences are implicitly assumed. Equation (4) is the IS schedule. The first term is the competitiveness effect; the second term is the real interest rate effect; the third term is the effect of higher activity in the foreign European country on the demand for home country output; the fourth term is the effect of fiscal policy; and the final term shows the shock to demand. The LM schedule is given by (5). The demand for nominal balances is a function of nominal income and the nominal interest rate. To simplify, I have assumed a unit nominal income elasticity of money demand.[5] Inflation is defined in terms of the simplest possible open-economy expectations-augmented Phillips-curve relationship (6). To prevent the order of dynamics from getting higher than what can be handled analytically, the "core" rate of inflation is assumed to equal the monetary growth rate and is therefore exogenous. The assumption of perfect capital mobility is reflected in terms of the uncovered interest rate parity condition (7) which closes the model. Finally, (8) and (9) define the real interest rate and the real exchange rate, respectively.

By eliminating the nondynamic variables, and before imposing any alternative policy rule, the equations (1)–(9) define a three-dimensional simultaneous linear differential equation system. For reasons of both analytical convenience and intuitive insight, I exploit Aoki's (1981) decomposition method. Thanks to the assumption of symmetry, the interdependent economy can be decomposed into two orthogonal subsystems: a (weighted) average system and a difference system. For simplicity, I also assume that the two countries have equal size, so any average, difference and single-country variable, respectively, can be written

$$x^a = \frac{1}{2}(x_1 + x_2); \quad x^d = x_1 - x_2; \quad x_i = x^a \pm \frac{1}{2}x^d; \quad i = 1, 2.$$

The average system describes a model for Europe as a whole (equivalent to a closed economy). The difference system is particularly useful for studying *convergence* between countries, an area of high priority both in the run-up to EMU and thereafter. Given the above assumptions, and by further assuming that the two countries have identical initial conditions, any difference in economic performance between the two countries must be attributed to asymmetric shocks.

The supply side is crucial for the discussion below. Two alternative (polar) structures are considered. In benchmark 1 I virtually neglect the supply side: there is no supply shock ($u = 0$) and constant returns to labor is assumed ($\beta \to \infty$). With a constant real product wage ($w = p$), firms are willing and able to accommodate any deviation of aggregate demand from its "natural" level. This benchmark serves to illustrate the potentials of having a strong endogenous shock absorber in the form of a relatively flexible labor market. In benchmark 2 the output elasticity of labor is less than one ($\beta < \infty$), there is full wage indexation ($\mu = 1$), and supply shocks may occur. This setting intends to resemble the current state of affairs in Europe, which is characterized by a relatively high degree of real wage rigidity (Bruno and Sachs, 1985; OECD, 1994). Subsequently, I refer to the two benchmarks as nominal wage rigidity (NWR) and real wage rigidity (RWR), respectively.[6]

While offering a transparent device for stabilization analysis, there are some potentially important restrictions implicit in the model. First, the order of dynamics is kept down by abstracting from *all* stock-flow issues and hence the model does not focus on budget deficits or government debt stocks. While this means that the model is appropriate only for short-run analysis, it may also have implications for the treatment of *permanent* shocks. For example, intuition would suggest that a permanent (negative) shock to aggregate demand would involve a permanent budget deficit and an increasing public debt. Since the model's restrictions do not allow this point to be analyzed, the policy implications derived from the model would be most relevant in the context of temporary shocks.

Second, the model does not allow for private sector co-insurance behavior. That is, if fiscal policy is rigid, it may still be possible for the private sector to borrow against the prospect of future adjustment. In models exhibiting consumption smoothing one would thus expect a weaker case for active fiscal policy (Masson and Melitz, 1991). However, while a micro-based, intertemporal approach would be more satisfactory from a theoretical point of view, the empirical evidence in favor of that approach is far from clear-cut.[7]

Third, the model can be criticized for using fixed parameters. There is no allowance for learning and adaptation, therefore. This restriction may be serious since some investigators view the formation of EMU as a regime change and the difference between flexible fiscal policy systems and rigid ones as a regime difference. If, for example, an economy is hit by a permanent supply shock, it could be argued that fiscal flexibility would limit the incentives to undertake the necessary real wage adjustments. The (relative) absence of fiscal policy as an alternative shock absorber may therefore not be such a bad thing: for it may hasten, not hinder, the development of more flexible real wage responses.

3. EMU in Europe

First consider the case of a fully-fledged monetary union in Europe. This regime is thought to resemble the blueprint for EMU, as envisaged by the founding fathers of the Delors Report (1989). I have in mind a Europe with one money (euro), managed by a common central bank (ECB) which strives to maintain price stability. The ECB is assumed to operate a flexible exchange rate *vis-à-vis* the rest of the world. The stock of foreign exchange reserves (f^a) is therefore constant and may, for simplicity, be set equal to zero ($f^a = 0$). The ECB's domestic credit expansion (d^a) is also assumed constant and equal to zero ($d^a = 0$), hence allowing for a stationary long-run equilibrium with price stability ($\pi = 0$).[8] Given the absence of risk and uncertainty, there can only be one interest rate and one LM schedule:

$$\bar{d}^a = p^a + y^a - \lambda i^a. \tag{10}$$

A rigid Maastricht interpretation would only allow fiscal policy to operate as a shock absorber if public finances are sufficiently strong. A softer interpretation, however, would assign a more active role to fiscal policy. Both interpretations may be encompassed within the following simple rule of fiscal policy:

$$s_i = \chi y_i; \quad \chi \geq 0, \quad i = 1, 2. \tag{11}$$

The model cannot demonstrate whether the required degree of fiscal flexibility actually exceeds what the Maastricht Treaty will allow. However, given the current debt and deficit ratios in Europe, it is evident that the fiscal criteria would be felt by nearly all EU countries. It should be noted, though, that the present European conjuncture may be exceptional. Also, the Maastricht Treaty numerical provisions are based on average behavior up to 1991, and the Treaty itself contains qualifying clauses on the role of these numerical provisions.[9]

Nominal Wage Rigidity

With demand-determined output, as obtained by neglecting equations (1)–(3), the two subsystems may be written on the following one-dimensional state–space forms:

$$\dot{p}^a = \frac{-\delta_2 \lambda^{-1} v}{\Delta + \Omega} p^a + \frac{v}{\Delta + \Omega} v^a, \tag{12}$$

$$\dot{p}^d = \frac{-2\delta_1 v}{\Delta - \Omega} p^d + \frac{v}{\Delta - \Omega} v^d, \tag{13}$$

where

$$\Delta \equiv 1 + \delta_2 \left(\frac{1}{2} \lambda^{-1} - v \right) + \delta_4 \chi \quad \text{and} \quad \Omega \equiv \frac{1}{2} \delta_2 \lambda^{-1} - \delta_3.$$

Since the price level is predetermined, both roots must be negative. In economic terms, these stability conditions ($\Delta \pm \Omega > 0$) say that the slope of the AD-curve in each subsystem must be nonpositive.

By appropriate fiscal action, both subsystems, and hence also each of the individual economies, may be protected against demand shocks: by letting χ tend to ∞, the coefficients $v/(\Delta \pm \Omega)$ drop out. In effect, the AD-curves become vertical. The story is simple: demand shocks spread through the system along the same transmission mechanism as that of fiscal policy and can therefore be redressed through fiscal feedback.

What are the costs (in terms of price and output deviations) of preventing fiscal policy from responding ($\chi = 0$)? Suppose the home country is hit by an immediate, unexpected, permanent and negative demand shock ($dv_1 < 0$), with $dv_2 = 0$. The real output effects are strictly short-run only, and the long-run price effects (denoted by "hats") read:

$$\frac{\partial \hat{p}_i}{\partial v_1} = \frac{2\delta_1 \pm \delta_2 \lambda^{-1}}{4\delta_1 \delta_2 \lambda^{-1}}; \quad i = 1, 2. \tag{14}$$

A permanent fall in home demand unambiguously leads to a lower home, average and difference price level, whereas the foreign price level may either rise or fall. To compensate for a loss in the exogenous component of aggregate demand, those parts sensitive to international competitiveness and the real interest rate must rise. This clearly requires a lower price level at home, both in absolute and relative terms. The shock is felt by the foreign country in the form of a lower real interest rate and a fall in international competitiveness. Whether the foreign price level also falls will depend on the relative strength of international competitiveness and the real interest rate as determinants of aggregate demand. If the former dominates (δ_1 "high"), the foreign price level must fall to partly offset the home country's gain in competitiveness.

The value of δ_1 is unlikely to be invariant to the stage of economic integration in Europe. Indeed, δ_1 may be interpreted as the degree of international integration of goods markets: the more closely integrated are home and foreign goods markets, the higher is δ_1. At an early EMU stage, including the run-up to EMU, δ_1 is likely to be fairly low. As economic integration progresses, δ_1 gradually becomes higher, with the "law of one price" eventually emerging ($p^d \to 0$ as $\delta_1 \to \infty$). With highly integrated goods markets, we would thus expect the home and foreign price levels to move in the same direction, irrespective of the nature and origin of the shock.

The dynamic behavior of the price level and output in both countries can easily be retrieved from the two subsystems:

$$p_i(t) = p_i(0) + \frac{1}{2}\left\{ \lambda\delta_2^{-1}\left[1 - \exp\left(-\frac{\delta_2\lambda^{-1}v}{\Delta' + \Omega}t \right) \right] \right.$$
$$\left. \pm \frac{1}{2}\delta_1^{-1}\left[1 - \exp\left(-\frac{2\delta_1 v}{\Delta' - \Omega}t \right) \right] \right\}dv; \quad i = 1,2 \tag{15}$$

$$y_i(t) = \frac{1}{2}\left\{ \frac{1}{\Delta' + \Omega}\exp\left(-\frac{\delta_2\lambda^{-1}v}{\Delta' + \Omega}t \right) \right.$$
$$\left. \pm \frac{1}{\Delta' - \Omega}\exp\left(-\frac{2\delta_1 v}{\Delta' - \Omega}t \right) \right\}dv; \quad i = 1,2, \tag{16}$$

where $\Delta' = \Delta - \delta_4\chi$. A negative shock to demand in the home country thus unambiguously causes a recession in that country (see Figure 1(a)). The price level gradually falls towards its new long-run level. Output, however, falls on impact before gradually returning to its natural rate level. The persistence of the home recession does depend on the degree of goods market integration: the higher is δ_1, the faster is the adjustment towards the new steady state. The magnitude of δ_1 is, of course, only relevant for the speed of adjustment of the difference system: the more closely integrated are the two economies, the faster is convergence achieved.

The overseas effects are not as clear-cut. They depend on the demand structure of the economy, notably the weight of competitiveness (δ_1) and the real interest rate (δ_2) as determinants of aggregate demand. Suppose $2\delta_1 > \delta_2\lambda^{-1}$, corresponding to highly integrated goods markets, or what could be a rather late stage of EMU. The impact effect of a slowdown in home output is a boost to foreign output (see Figure 1(b)). The reason for this rather counterintuitive result is that the real interest rate falls, thereby raising all interest-sensitive demand components. This effect comes about through a lower nominal interest rate, but since the price level adjusts only sluggishly, the real

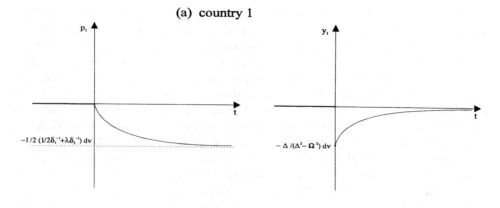

(a) country 1

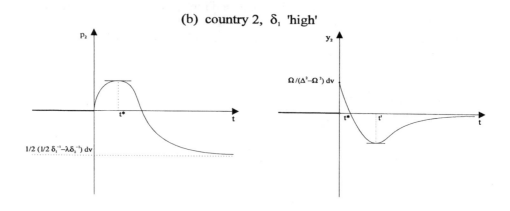

(b) country 2, δ_1 'high'

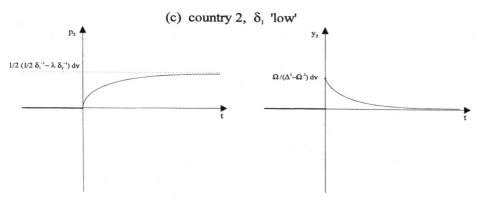

(c) country 2, δ_1 'low'

Figure 1. Regional Responses to Asymmeric Demand Shocks: EMU with Rigid Fiscal Policy

interest rate falls as well. Moreover, we observe that inflation in the two countries diverge for a period.[10] This happens until t^*, a time period which may indeed be very short. Thereafter, as competitiveness gradually deteriorates, country 2 will also be led into recession.

Suppose next $2\delta_1 < \delta_2\lambda^{-1}$, a situation likely to characterize an earlier stage of EMU. The overseas effects now depend on *how* high δ_1 is. If δ_1 is still relatively high, we get the same output performance as observed in Figure 1(b). If δ_1 is "very" low, however, the foreign price level will gradually increase towards its new long-run level, and country 2 experiences a boom throughout the adjustment process, as illustrated by Figure 1(c).[11] The price and output effects in country 2 are thus completely different from those in country 1.

This shows an important result, namely that it would be unwise to limit *both* monetary and fiscal responses. In particular, if goods markets are not highly integrated, as they might not yet be in the run-up to EMU or even at early stages of EMU, a simultanuous pursuit of rigid monetary and fiscal policy could seriously undermine the attempts to achieve convergence between European economies.[12] If a high weight is attributed to meeting the Maastricht convergence criteria, the results show that fiscal policy is needed: not only for *stabilization* purposes at Union level, but certainly also as a means to achieve *convergence* between members of the Union.

Real Wage Rigidity

Although often considered a workhorse for analyses of macroeconomic stabilisation policy, the above model may be an inadequate description of European economies. To account for the empirical evidence of a high degree of real wage rigidity in Europe, I now add the supply side under the extreme assumption of full wage indexation.[13]

I first use (1), (2), (3), and (9) to solve for the aggregate supply (AS) equation, and then I use (4), (6), (8), (9), (10), and (11) to solve for the aggregate demand (AD) equation. It turns out that all dynamics disappear in a monetary union with real wage rigidity in both countries. The AS schedule thus becomes vertical in the average system, so output is insulated against demand shocks at Union level. A supply shock, however, affects both the output level and the price level. The highest degree of average price stability would obviously be achieved by keeping the AD schedule horizontal.

While the potentials for fiscal policy at Union level are rather modest, some scope remains at the level of each member state. As is well-known, anything which alters the ratio of the real employer cost $(w - p)$ to the real consumption wage $(w - p^c)$ works like a supply shock. With $w - p^c$ held fixed, any stabilization must operate through the relative price level, $p^d (= -c)$. If goods markets are fully integrated, any incipient price difference is completely arbitraged away and hence demand shocks have no output effects even at regional level. Similar effects may be achieved by appropriate use of fiscal policy, even if goods are internationally differentiated. Suppose there is a negative shock to aggregate demand in country 1. Without fiscal feedback ($\chi = 0$), the relative price level would fall, thereby lowering output supply in that country, and a boom is exported to the foreign economy. However, it can easily be shown that the fall in relative prices can be avoided by appropriate fiscal stabilization. Fiscal policy may therefore influence the regional distribution of the price and output effects of asymmetric demand shocks.

Consider next the case where a supply shock hits country 1. The home and cross-country output and price effects read

$$\frac{\partial y_i}{\partial u_1} = -\frac{1}{2}\beta\left\{1 \pm \frac{\delta_1}{\beta\gamma(\Delta-\Omega)+\delta_1}\right\}; \quad i=1,2,\tag{17}$$

$$\frac{\partial p_i}{\partial u_1} = \frac{1}{2}\beta\left\{\lambda\delta_2^{-1}(\Delta+\Omega) \pm \frac{1}{2}\frac{\Delta-\Omega}{\beta\gamma(\Delta-\Omega)+\delta_1}\right\}; \quad i=1,2.\tag{18}$$

If there is no fiscal feedback, the home country immediately feels the shock in the form of lower output and higher prices. Although not directly affected, the foreign country may feel the shock indirectly. Again, the case with perfectly integrated goods markets constitutes an interesting benchmark: output adjustments would then entirely take place in the home country, whereas the costs in terms of higher prices are spread equally on the two countries. In the foreign country there is an equi-proportionate shift in the aggregate demand and aggregate supply schedules, hence leaving the foreign economy with a higher price level and output unaffected.

If the two countries face downward-sloping aggregate demand curves, however, there will be relative price effects, and the average fall in output will be spread between the two countries. The home supply shock will trigger an upward pressure on foreign nominal wages to protect real wages, so the foreign AS schedule will shift upwards. However, the rise in the foreign price level is smaller than the rise in the home price level, hence indirectly lowering foreign output supply.

The appropriate response to supply shocks depends on how the objective of nominal price stability is weighted relative to the objective of real output convergence. The fiscal authorities may keep the price level stable by adopting a "leaning with the wind" strategy ($\Delta - \Omega = 0$). Such a strategy implies, however, that all the burden of real adjustment will be placed on the country in which the shock occurs. Otherwise, by letting fiscal policy operate strongly "against the wind" ($\Delta - \Omega \rightarrow \infty$), a symmetric output response is feasible, but this solution is subject to the problem of indeterminacy of the price level.

4. A Case for More Flexibility in Monetary Policy?

In contrast to the Maastricht rules of rigid monetary policy discussed above, I now consider a more flexible framework for international coordination of monetary policy. The package examined in this section has much in common with Williamson and Miller's (1987) extended target zone proposal. It might be considered relevant in the run-up to EMU, and some would even see it as an appropriate long-term alternative to EMU! In particular, if strong constraints are imposed on fiscal policy during the transition to EMU, some flexibility in monetary policy could be desirable or even necessary for achieving convergence in other areas.

The rules envisaged for monetary policy involve a nominal income target and a real exchange rate target, both achieved through interest rate variations, corresponding to actual practice. Specifically, the *average* level of interest rates is set so as to achieve a target level of aggregate nominal income and the interest *differential* is set so as to maintain real exchange rates within certain limits:

$$i^a = \xi_1(p^a + y^a); \quad \xi_1 \geq 0,\tag{19}$$

$$i^d = \xi_2(e - p^d); \quad \xi \geq 0.\tag{20}$$

Equation (19) can be seen as a simplified version of the inverted LM curve (5), now with a possibility of speeding up the adjustment towards equilibrium through *policy*. Equation (20) is a rule of real exchange rate stabilisation. The "Keynesian" flavour of this proposal is reflected in the role it assigns for fiscal policy: since the conduct of monetary policy may result in an unsatisfactory macroeconomic performance, each country should be allowed to adjust fiscal policy in order to pursue a nominal income target

$$s_i = \xi_3\left(p_i + y_i\right); \quad \xi_3 \geq 0, \quad i = 1,2. \tag{21}$$

The value of ξ_3 will clearly depend on the strictness of interpretation of the fiscal convergence criteria.

Nominal Wage Rigidity

By solving the model under NWR, the two subsystems may be summarized on the following state–space forms:

$$\dot{p}^a = \frac{-v\left(\delta_2\xi_1 + \delta_4\xi_3\right)}{\Phi + \Psi} p^a + \frac{v}{\Phi + \Psi} v^a, \tag{22}$$

$$\begin{bmatrix} \dot{p}^d \\ \dot{e} \end{bmatrix} = \begin{bmatrix} \dfrac{-v\left(2\delta_1 + \delta_4\xi_3 - \delta_2\xi_2\right)}{\Phi - \Psi} & \dfrac{v\left(2\delta_1 - \delta_2\xi_2\right)}{\Phi - \Psi} \\ -\xi_2 & \xi_2 \end{bmatrix} \begin{bmatrix} p^d \\ e \end{bmatrix} + \begin{bmatrix} \dfrac{v}{\Phi - \Psi} \\ 0 \end{bmatrix} v^d \tag{23}$$

where

$$\Phi \equiv 1 + \delta_4\xi_3 + \delta_2\left(\frac{1}{2}\xi_1 - v\right) \quad \text{and} \quad \Psi \equiv \frac{1}{2}\delta_2\xi_1 - \delta_3.$$

Stability of the average system (22) again implies a nonpositively sloped AD-curve ($\Phi + \Psi > 0$). As to the assignment of monetary and fiscal policy, I observe that at Union level an active pursuit of a nominal income target can insulate against demand shocks. Such an outcome may be achieved through either coordinated monetary policy ($\xi_1 \to \infty$), or by letting each individual country adopt such a target through fiscal policy ($\xi_3 \to \infty$). In either case, the AD schedule becomes immune to demand shocks.

As to the difference system (23), a saddlepoint equilibrium requires that the characteristic roots of the homogenous system carry different sign, that is the determinant of the state matrix, $\Delta_A = -\xi_2 v \delta_4 \xi_3/(\Phi - \Psi)$, must be negative. Again, this requires a nonpositively sloped AD curve. Also, *some* fiscal feedback is necessary ($\xi_3 > 0$). The point is that in policy regimes with real exchange rate targets, monetary policy *alone* provides no anchor for domestic inflation: a given real exchange rate is consistent, in principle, with any evolution of nominal variables (Adams and Gros, 1986). In the complete absence of fiscal feedback ($\xi_3 = 0$), the new steady-state price differential will tend to infinity. However, with fiscal policy being used in conjunction with real exchange rate targeting, one should not worry about this indeterminacy problem.

I now turn to examine individual country performance given a negative demand shock in country 1. Prices and outputs are governed by the following equations:

$$p_i(t) = p_i(0) + \frac{1}{2} \left\{ \frac{1}{\delta_2 \xi_1 + \delta_4 \xi_3} \left(1 - \exp\left(- \frac{v(\delta_2 \xi_1 + \delta_4 \xi_3)}{\Phi + \Psi} t \right) \right) \right.$$

$$\left. \pm \frac{1}{\delta_4 \xi_3} \left(1 - \exp(\rho t) \right) \right\} dv; \quad i = 1,2, \tag{24}$$

$$y_i(t) = \frac{1}{2} \left\{ \frac{1}{\Phi + \Psi} \exp\left(- \frac{v(\delta_2 \xi_1 + \delta_4 \xi_3)}{\Phi + \Psi} t \right) \pm \frac{\rho}{v \delta_4 \xi_3} \exp(\rho t) \right\} dv;$$

$$i = 1,2, \tag{25}$$

where ρ, the stable root in the difference system, is given by

$$\rho = \frac{1}{2(\Phi - \Psi)} \left\{ (\Phi - \Psi)\xi_2 - v(2\delta_1 + \delta_4 \xi_3 - \delta_2 \xi_2) \right.$$

$$\left. - \sqrt{\left[(\Phi - \Psi)\xi_2 - v(2\delta_1 + \delta_4 \xi_3 - \delta_2 \xi_2) \right]^2 + 4(\Phi - \Psi)\xi_2 v \delta_4 \xi_3} \right\}.$$

Given the complexity of ρ, a purely algebraic treatment of price and output behavior is a difficult task.[14] However, two important results can be offered. First, by appropriate fiscal design ($\xi_3 \to \infty$), all price and output effects of asymmetric demand shocks can be nullified. This is not a great deal, of course, as this can be achieved even in a monetary union. Second, while a demand shock can be redressed at Union level by adhering to a strict nominal income rule, the regions cannot be completely protected. Monetary policy can thus not fully make up for the lack of fiscal flexibility. This is a strong result, pointing to the dangers of limiting fiscal responses, even if some monetary flexibility is available.

Figure 2 shows the regional adjustment trajectories of output and the price level following a demand shock in country 1, assuming that fiscal policy is rigid. To obtain comparability with Figure 1, cases are illustrated with alternative degrees of international integration of goods markets. Qualitatively, the EMU results go through, that is, country 1 is led into recession, while the spillovers depend on, inter alia, the degree of goods market integration: in weakly integrated economies asymmetric demand shocks may give rise to serious convergence problems.

These results may be further developed with the aid of some numerical illustrations. Think of four performance criteria: first, the initial jump in output, $y(0)$; second, the max/min value of output during the adjustment period, $y(m)$; third, the price level in the new steady state, $p(\infty)$; and fourth, the speed of adjustment towards the new steady state, $\tilde{\rho}$. Given a "sensible" set of structural parameter values, these criteria are applied to both the average and the difference systems, and to both regions. Since our focus is on the role of monetary flexibility, fiscal policy is considered to be "rigid."[15] As to monetary policy, two alternatives are considered: in scenario A, ξ_1 is set equal to the inverse of the interest rate semielasticity of the demand for money ($\xi_1 = \lambda^{-1} = 1.33$), while scenario B assumes a higher coefficient of response ($\xi_1 = 10$). In both scenarios a relatively strict real exchange rate rule is adopted ($\xi_2 = 20$).[16]

The results are summarized in Table 1. Two questions are relevant: how does a flexible monetary regime (I) perform in comparison with a monetary union (II), and

(a) country 1

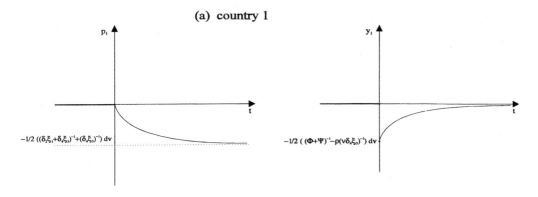

(b) country 2 , δ_1 'high'

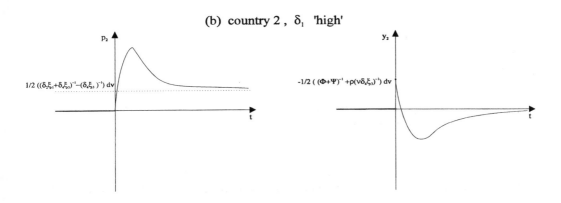

(c) country 2, δ_1 'low'

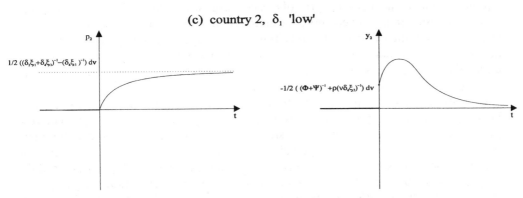

Figure 2. Regional Responses to Asymmetric Demand Shocks: Flexible Monetary Policy and Rigid Fiscal Policy

Table 1. Stabilizing Function of Monetary Policy Under Nominal Wage Rigidity

(I) *Monetary flexibility*

	Average		Difference		Country 1		Country 2	
	A	B	A	B	A	B	A	B
$dv = -1$								
$y(0)$	-0.29	-0.05	-0.74	-0.74	-0.67	-0.42	0.08	0.32
$y(m)$	-0.29	-0.05	-0.74	-0.74	-0.67	-0.42	-0.0018	0.37
$p(\infty)$	-0.37	-0.05	-200	-200	-100.4	-100	99.6	100
ρ	-0.19	-0.24	-0.00093	-0.00093	—	—	—	—

(II) *EMU*

	Average	Difference	Country 1	Country 2
$dv = -1$				
$y(0)$	-0.29	-0.91	-0.74	-0.17
$y(m)$	-0.29	-0.91	-0.74	-0.0015
$p(\infty)$	-0.36	-1	-0.88	-0.13
ρ	-0.19	-0.23	—	—

Parameter values: $\delta_1 = 0.5$, $\delta_2 = 1$, $\delta_3 = 0.35$, $\delta_4 = 0.5$, $\lambda = 0.75$, $v = 0.25$, $\chi = 0.00$.
A: $\xi_1 = 1.33$, $\xi_2 = 20$, $\xi_3 = 0.01$; B: $\xi_1 = 10$, $\xi_2 = 20$, $\xi_3 = 0.01$.

what difference does it make whether a "soft"(A) or a "tough"(B) nominal income rule is pursued at Union level?

The comparison of the two monetary regimes shows that, as long as the response coefficient equals the (inverse of the) interest elasticity, scenario A, it makes no difference for the average economy which of the two regimes is operated. It also turns out that on impact the highest degree of *real* convergence is achieved by the target zone regime, a result that is even more pronounced if goods markets are only weakly integrated.[17] On the other hand, *nominal* price convergence is much more effectively delivered in a monetary union. Indeed, unless supported by strong fiscal control, a target zone regime leads to large price differences and the new steady state is reached only extremely slowly.

How strongly should the average nominal income target be adhered to? While a higher coefficient of response dampens the macro effects at Union level and speeds up the adjustment towards the new steady state (with complete insulation being offered in the limiting case of $\xi_1 \to \infty$), the value of ξ_1 is completely irrelevant for convergence, as shown by the difference system. At regional level, a higher value of ξ_1 means that while country 1's initial output loss is smaller, the rise in country 2's output is correspondingly higher, with the difference between them left unchanged. Not only is the foreign output effect stronger on impact, the (nonmonotonic) output profile in country 2 shows that output even continues to increase after the initial shock, before returning to the initial level.

In sum, without fiscal feedback, the two monetary regimes really offer a tradeoff to policymakers: if they have a strong priority for nominal convergence, as in the

Maastricht Treaty, a monetary union should be opted for, just as the Maastricht Treaty does. If, instead, they are more for real convergence, a criterion hardly mentioned in the Maastricht Treaty, they should side with the target zone proposal.

Real Wage Rigidity

Unlike the case of real wage rigidity under EMU, real wage rigidity with monetary flexibility produces some genuine exchange rate dynamics, as shown by the following differential equation derived from the difference system:

$$\dot{e} = \frac{1}{\Lambda}\left\{\xi_2\delta_4\xi_3 e - \xi_2 v^d - \xi_2\beta(\Phi - \Psi)u^d\right\}, \tag{26}$$

where

$$\Lambda \equiv 2\delta_1 - \delta_2\xi_2 + \delta_4\xi_3 + 2\gamma\beta(\Phi - \Psi).$$

In the following I assume that the coefficient of response (ξ_2) is sufficiently high so as to ensure a stable relationship.

An important difference is that with monetary flexibility there is a possibility of redressing the price effects of demand shocks, so that these may not be felt at Union level.

We now turn to the regional price and output responses, which may be written as follows:

$$p_i = \frac{1}{2}\left\{\frac{1}{\delta_2\xi_1 + \delta_4\xi_3} \pm \frac{1}{\delta_4\xi_3}\right\}dv + \frac{1}{2}\left\{\frac{\beta(\Phi + \Psi)}{\delta_2\xi_1 + \delta_4\xi_3} \pm \frac{\beta(\Phi - \Psi)}{\delta_4\xi_3}\right\}du$$

$$\mp \frac{dv + \beta(\Phi - \Psi)du}{2\delta_4\xi_3}\exp\left(\frac{\xi_2\delta_4\xi_3}{\Lambda}t\right); \quad i = 1,2, \tag{27}$$

$$y_i = \pm \frac{\beta\gamma dv}{\Pi}\exp\left(\frac{\xi_2\delta_4\xi_3}{\Lambda}t\right)$$

$$-\frac{1}{2}\beta\left\{1 \pm \left[1 - \frac{2\gamma\beta(\Phi - \Psi)}{\Pi}\exp\left(\frac{\xi_2\delta_4\xi_3}{\Lambda}t\right)\right]\right\}du; \quad i = 1,2, \tag{28}$$

where $\Pi = 2\delta_1 - \delta_2\xi_2 + 2\gamma\beta(\Phi - \Psi)$. It is easily verified that fiscal policy may be designed so as to nullify the influences of demand shocks, and that it may contribute to speed up the process of convergence in response to supply shocks. The question is what would happen if fiscal policy is prevented from serving these purposes.

Again, the discussion may be illuminated by some numerical examples. I use the same parametrization as before, now extended with the two supply-side parameters, β and γ.[18] The results are reported in Table 2.

In response to demand shocks, the two regimes perform identically as long as $\xi_1 = \lambda^{-1}$. However, by sticking more firmly to the nominal income target (scenario B), the Union as a whole may do better. Notice that since ξ_1 is not relevant for the difference system, and since only the difference system is actually dynamic, neither the differences nor the regional performances depend on how strongly the average nominal income target is pursued.

Table 2. Stabilizing Function of Monetary Policy Under Real Wage Rigidity

(I) *Monetary flexibility*

	Average		Difference		Country 1		Country 2	
	A	B	A	B	A	B	A	B
$dv = -1$								
$y(0)$	0	0	0.18	0.18	0.088	0.088	−0.088	−0.088
$p(\infty)$	−0.37	−0.05	−200	−200	−100.4	−100.1	99.6	100
ρ	—	—	−0.0063	−0.0063	−0.0063	−0.0063	−0.0063	−0.0063
$du = 1$								
$y(0)$	−2	−2	−4.78	−4.78	−4.39	−4.39	0.39	0.39
$p(\infty)$	2.6	2.1	884	884	444.6	444.1	−439.4	−439.9
ρ	—	—	−0.0063	−0.0063	−0.0063	−0.0063	−0.0063	−0.0063

(II) *EMU*

	Average	Difference	Country 1	Country 2
$dv = -1$				
y	0	−0.69	−0.34	0.34
p	−0.38	−0.25	−0.49	−0.25
$du = 1$				
y	−2	−0.98	−2.49	−1.51
p	2.6	1.08	3.14	2.06

Parameter values: $\delta_1 = 0.5$, $\delta_2 = 1$, $\delta_3 = 0.35$, $\delta_4 = 0.5$, $\lambda = 0.75$, $v = 0.25$, $\beta = 4$, $\gamma = 0.35$, $\chi = 0.00$.
A: $\xi_1 = 1.33$, $\xi_2 = 20$, $\xi_3 = 0.01$; B: $\xi_1 = 10$, $\xi_2 = 20$, $\xi_3 = 0.01$.

I also recognize the tradeoff between nominal and real convergence: without fiscal control the target zone proposal fails to pin down prices and ensure convergence of price levels. On the other hand, it certainly outperforms the monetary union when it comes to real convergence.

The most interesting results are related to the regional responses. I note that by allowing for some exchange rate flexibility, the output effects are certainly dampened. Not only does the quantitative significance differ, even the sign of response is different across regimes. Consider first a negative demand shock in country 1. That leads to an appreciation of the nominal exchange rate (see (26)) and, with predetermined prices, also to real appreciation. This clearly boosts country 1's output supply, at the expense of country 2's output, so that the overall Union effects wash out. On the other hand, since the real exchange rate is managed by the monetary authorities, a cornerstone of the target zone proposal, the output effect is relatively minor. After some slow adjustment, output returns to its initial steady-state level and the countries' price levels diverge.

Consider next a supply shock. Apparently, a target zone system results in a stronger impact effect on output in country 1 than would be the case under EMU. While this outcome is not robust, it would typically be the case if a real exchange rate target is strictly adhered to.[19] The reason is as follows. Irrespective of the monetary regime, there will first be an *exogenous* fall in output when the wage shock appears. In a second round, there will be some *endogenous* output responses. Under the target zone system, the stability condition dictates a depreciation of the exchange rate (cf. (26)). With real wage rigidity, the depreciation would cause a rise in the real employer cost relative to the real consumption wage. This "wedge" works like another adverse supply shock, which brings further downward pressure on output. Clearly, this effect would not be present in a monetary union.

5. Conclusion

Asymmetric shocks are likely to remain important, also *after* the formation of a monetary union. Their occurrence may seriously challenge the sustainability of the EMU project, perhaps eventually ruining it. But the macroeconomic impact of asymmetric shocks does depend on both the economic structure and the policy framework. The labor market is of particular importance, and macroeconomic policy should be designed in recognition of prevailing structures.

The analysis conveys one very clear message: flexibility in fiscal policy is a necessary ingredient in a policy package for EMU. As long as the economy is predominantly hit by demand shocks, fiscal policy can fully counter such shocks. However, even if real wages are relatively flexible, and hence the endogenous shock absorbers are strong, asymmetric shocks may threaten macroeconomic stability: fiscal policy can speed up the stabilization process, at both Union and regional level. If real wages are rigid, as they typically are in Europe, fiscal policy cannot remove the effects of asymmetric supply shocks, but it can successfully contribute to achieving convergence between member states.

Fiscal policy cannot replace labor market reforms and other supply side measures. But it would be dangerous to give up fiscal flexibility at the same time as monetary autonomy is diminished. On the other hand, monetary flexibility, a possible option in the run-up to EMU, cannot completely make up for the stabilization function of fiscal policy. Indeed, as long as fiscal policy is flexible, this analysis makes no case against monetary union. On the contrary: not only does monetary union perform very well in conjunction with fiscal policy, it also automatically secures against financial shocks, which are likely to appear frequently in regimes with monetary flexibility.

References

Adams, C. and D. Gros, "The Consequences of Real Exchange Rate Rules for Inflation: Some Illustrative Examples," *IMF Staff Papers* 33 (1986):439–76.

Alogoskoufis, G., "Stabilization Policy, Fixed Exchange Rates and Target Zones," in M. Miller, B. Eichengreen, and R. Portes (eds.), *Blueprints for Exchange Rate Management*, New York: Academic Press, 1989.

Aoki, M., *Dynamic Analysis of Open Economies*, New York: Academic Press, 1981.

Argy, V. and J. Salop, "Price and Output Effects of Monetary and Fiscal Expansion in a Two-Country World Under Flexible Exchange Rates," *Oxford Economic Papers* 13 (1983):228–46.

Bayoumi, T. and B. Eichengreen, "Shocking Aspects of European Monetary Unification," in F. Torres and F. Giavazzi (eds.), *Adjustment and Growth in the European Monetary Union*, Cambridge: Cambridge University Press, 1993.

Bayoumi, T. and E. Prassad, "Currency Unions, Economic Fluctuations and Adjustment: Some Empirical Evidence," CEPR discussion paper 1172, 1995.

Bruno, M. and J. Sachs, *The Economics of Worldwide Stagflation*, Oxford: Basil Blackwell, 1985.

Buiter, W., "Macroeconomic Policy Design in an Interdependent World Economy: An Analysis of Three Contingencies," in J. Frenkel (ed.), *International Aspects of Fiscal Policies*, Chicago: Chicago University Press, 1988.

Buiter, W., "Macroeconomic Policy During a Transition to Monetary Union," CEPR discussion paper 1222, 1995.

Buiter, W., G. Corsetti, and N. Roubini, "Excessive Deficits: Sense and Nonsense in the Treaty of Maastricht," *Economic Policy* 16 (1993):57–101.

Christodoulakis, N., S. Dimelis, and T. Kollintzas, "Comparisons of Business Cycles in the EC: Idiosyncracies and Regularities," *Economica* 62 (1995):1–27.

Delors Committee, "Report on Economic and Monetary Union in the European Community," Office for Official Publications of the European Community, Luxembourg, 1989.

Dornbusch, R., "Expectations and Exchange Rate Dynamics," *Journal of Political Economy* 84 (1976):1161–76.

European Commission, "One Market, One Money," *European Economy* 44, 1990.

Hughes Hallett, A. and P. McAdam, "Fiscal Deficit Reduction in Line with the Maastricht Criteria for Monetary Union: An Empirical Analysis," CEPR discussion paper 1351, 1996.

Hughes Hallett, A. and D. Vines, "On the Possible Costs of European Monetary Union," *The Manchester School* 61 (1993):35–64.

Jensen, S.H. and L.G. Jensen, "Debt, Deficits and Transition to EMU: A Small Country Analysis," *European Journal of Political Economy* 11 (1995):3–25.

Manasse, P., "Fiscal Policy in Europe: The Credibility Implications of Real Wage Rigidity," *Oxford Economic Papers* 32 (1991):1031–54.

Masson, P. and J. Melitz, "Fiscal Policy Independence in a European Monetary Union," *Open Economies Review* 2 (1991):113–36.

Meade, J. and M. Weale, "Monetary Union and the Assignment Problem," *Scandinavian Journal of Economics* 97 (1995):201–22.

Miller, M. and A. Sutherland, "The Walters Critique of the EMS—A Case of Inconsistent Expectations," *Manchester School* 59 (1991):23–37.

Miller, M. and J. Williamson, "The International Monetary System: An Analysis of Alternative Regimes," *European Economic Review* 32 (1988):1031–54.

OECD, "The OECD Jobs Study: Facts, Analyses, Strategies," Paris, 1994.

Turnovsky, S., "Monetary and Fiscal Policy Under Perfect Foresight: A Symmetric Two-Country Analysis," *Economica* 53 (1985):139–57.

Von Hagen, J. and S. Lutz, "Fiscal and Monetary Policies on the Way to EMU," *Open Economies Review* forthcoming.

Walters, A., *Sterling in Danger*, London: Fontana, 1990.

Williamson, J. and M. Miller, *Targets and Indicators: A Blueprint for the International Coordination of Economic Policy*, Washington, DC: Institute for International Economics, 1987.

Notes

1. See Buiter et al. (1993) for a lengthy and critical appraisal of the Maastricht fiscal convergence criteria. The interactions between monetary and fiscal policy in a monetary union (and in the run-up to it) have also been critically assessed by Allsopp et al. (this issue), Buiter (1995), Von Hagen and Lutz (1996), Hughes Hallett and McAdam (1996), Hughes Hallett and Vines (1993), Jensen and Jensen (1995), and Meade and Weale (1995).

2. The proposal we have in mind is that of Williamson and Miller (1987). It received a lot of attention in the late 1980s and was by many seen as an appropriate blueprint for international coordination of economic policy at the G-3 level. See also Miller and Williamson (1988) and Alogoskoufis (1989).

3. The model is basically set up *ad modum* Dornbusch (1976). Two-country extensions of that model can be found in Turnovsky (1985) and Buiter (1988), but in their models output is entirely

demand determined. Within a model for a small open economy, Manasse (1991) incorporates a supply side similar to ours.

4. The shocks are stylized, like the model in general, and empirically relevant magnitudes cannot be represented.

5. By deflating both the nominal money stock and nominal income by the consumer price index $(m - p^c = \phi(p + y - p^c) - \lambda i)$, and by setting the real income elasticity of real money demand to one $(\phi = 1)$, we obtain (5).

6. This emphasis of polar cases is common in the literature (Argy and Salop, 1983).

7. Indeed, two recent empirical studies—Von Hagen and Lutz (1996) and Hughes Hallett and McAdam (1996)—suggest that the contractionary effects of fiscal reductions in line with the Maastricht criteria would be quite strong.

8. In principle, and as a "soft" interpretation of what the ECB actually can do, the ECB could choose a nominal income target. In this section, however, we restrict attention to an EMU setting where monetary policy targets the money supply. Nominal income targeting is considered in the next section.

9. It should also be noted that, in practice, the rule (11) would typically be asymmetric: while a country with a debt ratio above the norm would be forced into fiscal reductions, it would hardly be realistic to bring pressure on a surplus country to spend more and thus reduce its surplus beyond its own will. Obviously, asymmetric rules that limit only deficits can lead national budgetary policies to be too restrictive in the aggregate.

10. This effect can be seen as a mild version of the so-called "Walters Critique," named after Mrs Thatcher's personal economic adviser, Sir Alan Walters. The basic claim of this critique is that it will not be possible to produce inflation convergence under fixed exchange rates: in the absence of expected exchange rate changes, there will be no interest differential and this will cause inflation rates to diverge. Walters himself has stated the critique as follows: "If Italy is inflating at a rate of 7 per cent and Germany at a rate of 2 per cent (both over the relevant period of exchange rate fixing), then there is a problem of perversity. With the same interest rate at 5 per cent, the real interest rate for Italy is minus 2 per cent and for Germany plus 3 per cent. Thus Italy will have an expansionary monetary policy, while Germany will pursue one of restraint. But this will exacerbate inflation in Italy and yet restrain further the already low inflation in Germany. This is the opposite of 'convergence', namely it induces divergence" (Walters, 1990, pp. 79–80). For a critical appraisal of the Walters Critique, see Miller and Sutherland (1991).

11. Formally, this result holds if $2\delta_1 v/(\Delta - \Omega) - \delta_2 \lambda^{-1} v/(\Delta + \Omega) \leq 0$.

12. It should be noticed that the monetary union considered in this section is qualitatively similar to a regime with irrevocably fixed exchange rates. Therefore, it can also be seen as a "hard-EMS": the monetary regime that, in the run-up to EMU, is more or less implied by the Maastricht provisions.

13. Real wage rigidity is here assumed to be symmetric. In Europe, however, it would be most realistic to think of real wages being rigid downwards but not upwards. Shocks causing lower prices may therefore not put sufficiently strong downward pressure on nominal wages as assumed by (2).

14. It is readily seen that $\rho = 0$ if $\xi_3 = 0$, a hysteresis property that explains the price level indeterminacy problem alluded to above.

15. Since the target zone regime requires some "minimal" fiscal control, we set $\xi_3 = 0.01$. Under EMU, we set $\chi = 0$.

16. If the real exchange rate is kept constant $(\xi_2 \to \infty)$, the interest rate differential should simply be set equal to the anticipated inflation differential so that real interest rates become equal $(i^d = \dot{p}^d = vy^d)$. Relative output demand would then only depend on differences in fiscal policy and asymmetric shocks.

17. The sensitivity of these criteria to the degree of international integration of goods markets is not reported in Table 1.

18. The wage share in production is assumed to equal 80%, and the share of import prices in the

CPI calculation is 35%. We have skipped the max/min value, as the adjustment trajectories are now clearly monotonic.

19. Formally, we have

$$\left(\frac{dy_1}{du_1}\right)_{emu} = -\frac{1}{2}\beta\left\{1+\frac{\delta_1}{\beta\gamma(\Delta-\Omega)+\delta_1}\right\} \underset{<}{\overset{>}{=}} \left(\frac{dy_1(0)}{du_1}\right)_{flex}$$

$$= \frac{1}{2}\beta\left\{1+\left[1-\frac{2\gamma\beta(\Phi-\Psi)}{2\delta_1-\delta_2\xi_2+2\gamma\beta(\Phi-\Psi)}\right]\right\}.$$

Review of International Economics, Special Supplement, 55–76, 1997

Monetary and Fiscal Stabilization of Demand Shocks Within Europe

*Chris Allsopp, Gareth Davies, Warwick McKibbin, and David Vines**

Abstract

This paper examines alternative macroeconomic stabilization rules for demand shocks, for a single open economy, and for an integrated European region. These questions are tackled in two ways. First a very simple macroeconomic model is used to focus on intercountry interconnections. Then the effects of shocks are simulated using the McKibbin Sachs MSG2 global economic model. The theoretical model analyzes just how much larger the disturbances caused by asymmetric shocks might be in a European Monetary Union, as compared with outcomes under floating exchange rates, especially (1) if rigid central monitoring and discipline of fiscal policy prevents the full operation of the inbuilt fiscal stabilizers within individual European countries, and (2) if European monetary policy does not concern itself with fully European objectives. Simulations with the MSG2 model bear out the significance of these risks. They show that a demand shock like GEMU can have strongly negative effects on output in other European countries if either interest rates are raised to counter the demand shock in the originating country, or if, for some reason, fiscal stabilization is not allowed to be as strong as the inbuilt fiscal stabilizers.

1. Introduction

This paper examines alternative macroeconomic stabilization rules for demand shocks, for an integrated European region. More generally it is about the design of the appropriate interconnections between monetary and fiscal policy within a European Monetary Union (EMU). We tackle these questions in two ways. First we use a very simple macroeconomic model to focus on intercountry interconnections. Then we use the insights from the simple model to undertake simulations using the European version of the McKibbin–Sachs MSG2 global economic model.

We first construct a simple IS–LM–BP macroeconomic model to demonstrate the familiar result that the stabilization of demand shocks under a floating exchange rate is greater than that under a fixed exchange rate. We then extend our model to a two-country region ("Europe") located in a large world. We show that, in a two-country region, positively correlated demand shocks are "bottled up" and magnified, augmenting each other through positive demand spillovers between the countries. Conversely, negatively correlated shocks are dissipated and damped, because positive demand spillovers between the countries tend to dampen regional differences.

Our simple model enables us to describe the macroeconomic policy properties of EMU in the following simple way. It has two components. The *stronger* shock-absorbing properties of floating exchange rates are deployed against the effects of demand shocks that are correlated between the countries, shocks which are anyway stronger because of the inbuilt intercountry spillovers which magnify them. By contrast merely the *weaker* shock-absorbing properties of fiscal feedbacks are deployed against negatively correlated demand shocks, which are already weaker because the

* Allsopp: New College, Oxford. Davies: Nuffield College, Oxford. McKibbin: Australian National University and Brookings Institution. Vines: Balliol College, Oxford, Australian National University, and CEPR. Tel: (44) 1865 271067; Fax: (44) 1865 271094; Email: econdav@ermine.ox.ac.uk.

inbuilt intercountry spillovers dampen them. These could be operated by national governments or by a central fiscal authority in Brussels.

One way of understanding the working of these two components is to examine the conditions required so that a positive demand shock in one country does not have a negative effect on output of the other country: these are that the positive demand spillover from the first country's output to the second country's output outweigh the damping effect on the second country's output of the currency appreciation of Europe as a whole.

Our simple model enables us to illuminate two design-risks to such an EMU system.

The first fear is that system-design problems on the fiscal side will lead to the weakening of the fiscal feedbacks operating within in each country. The original strict Maastricht fiscal conditions would have had this effect, in that they would have left no room for the operation of counter-cyclical stabilization at the level of the individual country (for fear that an inability to monitor whether fiscal deficits really were merely counter-cyclical would leave a loophole for countries to pursue structurally excess deficits). Such a weakening of fiscal feedbacks can be shown to increase the likelihood that a positive demand shock in one country leads to an output fall in the other country, because it increases the burden of the shock absorption which would be borne by the Europe-wide currency appreciation, and because it weakens the extent to which the fiscal policies in both countries dampen the disparities between them which emerge from the shock.

The second fear arises from the "Kenen problem" (Kenen, 1992). System design problems on the monetary side might leave monetary restraint for Europe as a whole responding to demand disturbance in just one region. This was the problem with the ERM in response to GEMU. We formally analyze a system of this kind in our simple model—we call it "ERM." (It is a "hard-ERM": we do not allow intercountry exchange rate differences in such a hard ERM regime.) A regime of this kind can be shown to increase the likelihood that a positive demand shock in one region will lead to an output fall in the other region, because the monetary restraint in response to the shock—and so the Europe-wide currency appreciation—is greater.

We investigate these questions in the light of the two shocks that have posed the greatest threat to European monetary integration in the last 10 years: the UK consumption boom of the late 1980s and the German reunification (GEMU) shock of the early 1990s. We simulate the effects of these shocks using the MSG2 multicountry economic model, a multicountry dynamic general equilibrium model of the world economy. First we simulate a UK consumption shock, approximately the size of the UK spending boom of the late 1980s, under different policy assumptions: (1) floating exchange rates; and (2) fixed exchange rates with three alternative strengths of fiscal feedback rules—the inbuilt stabilizers operating, the inbuilt stabilizers switched off, and the inbuilt stabilizers strengthened. We also simulate a fiscal shock in Germany, approximately the size of the German reunification shock of the early 1990s, under different policy assumptions for Europe as a whole: (1) floating exchange rates; (2) EMU with the same three alternative fiscal feedback rules, and (3) ERM with again the same three alternative fiscal feedback rules. Results are displayed for Germany and for the UK as a representative non-German European economy. The results illustrate and confirm the issues raised by our simple theoretical model.

Our paper is designed to contribute to the debate about European monetary unification, by displaying different alternative policy responses to demand shocks. We analyse only demand shocks because the dangers of responding to supply shocks through demand stabilization policy are well known. The model and the results provide an attempt to analyze and quantify just how much larger the disturbances caused by asymmetric shocks might be in a European Monetary Union, as compared with outcomes under floating exchange rates, especially (1) if rigid central monitoring and discipline of fiscal policy prevents the full operation of the inbuilt fiscal stabilizers within individual European countries, and (2) if European monetary policy does not concern itself with fully European objectives.

In more detail, the paper is laid out as follows. In section 2 we set up simple macroeconomic model of a single open economy, and then extend our analysis to a two-region model of "Europe"; the model is entirely demand-determined consisting only of an IS curve and LM curve and exchange rate determination (a BP curve). In section 3 we briefly describe the MSG2 model. Section 4 considers the effects of a demand shock in our single open-economy model, and simulates the counterpart empirical shock—a UK consumption shock—using the MSG2 model. Section 5 considers the effect of a range of symmetric and asymmetric demand shocks within Europe and simulates a counterpart empirical shock—the GEMU shock—using MSG2.

2. An IS–LM–BP Model of Fiscal and Monetary Policy in Europe

An Open Economy Model

In this section we will present a model of one single European economy, but identify influences on it from the other European country. All variables are treated as differences from a pre-shock equilibrium, so a value of zero simply means "no change." All shocks will be treated as temporary, and there is no explicit treatment of adjustment dynamics, although that will play an important part in the simulations which follow.

For convenience we now define the symbols which we will use in what follows. Starred variables, where they appear, are those for the foreign European country. Variables for the rest of the world ("the USA") are all set to zero.

y = level of domestic final production, measured in proportional deviations from its initial level

y^* = level of final production within the foreign European economy, again measured in proportional deviations from its initial level

r = domestic nominal interest rate, measured in percentage points

r^* = the nominal interest rate in the foreign European economy, measured in percentage points

e = the nominal exchange rate of the domestic economy *vis-à-vis* the US dollar, defined as the domestic currency price of dollars, so that an increase in e denotes a real depreciation, and measured as percentage deviations from initial level

e^* = the nominal exchange rate of the foreign European economy *vis-à-vis* the USA, defined according to the same convention, and measured in the same way as e

v = a shock to domestic demand.

The IS curve   Output is determined by demand, according to an open economy IS curve:

$$y = \delta e + \sigma(e - e^*) - \gamma r + \phi(1 - \tau)y - \mu y + \eta y^* + v. \tag{1}$$

The first two terms are competitiveness effects. These come in two parts, the first depending on competitiveness *vis-à-vis* the US, e, and the second depending on competitiveness *vis-à-vis* the foreign European country, $e - e^*$. The coefficients on these two competitiveness terms depend on the price elasticity of demand for exports, on the openness of the economy, and on the relative importance of intra-European trade versus non-European trade, but the details of this need not concern us here. The next term shows the effects on demand of domestic interest rates. The following term shows consumption effects, which depends on post-tax income, $(1 - \tau)y$. (We denote the tax rate by the parameter τ.) The import leakage is denoted by μy, and the term ηy^* shows the effects of higher activity in the foreign European country on the demand for home country output. The final term shows the shock to domestic demand.

Monetary and fiscal policy   We now consider the setting of macroeconomic policy instruments. In this paper we will examine the effects of simple fiscal and monetary rules. Interest rate determination is shown in equation (2), which is a form of LM curve:

$$r = \beta y. \tag{2}$$

With $\beta > 0$, this shows the conduct of monetary policy under floating exchange rate regime with a fixed money supply, implying that the interest rate rises whenever the nominal income increases (which in a fixed-price model means whenever output increases).[1] An alternative interpretation of this rule is that it is an explicit nominal income rule, in which the nominal interest rate is raised whenever nominal income is above target (Taylor, 1994).[2] Under fixed exchange rates, $\beta = 0$; home interest rates are pegged to world (US) interest rates.

We need also to specify the responsiveness of fiscal policy to changes in domestic activity, through means of the responsiveness of taxes to output. We will examine the effects of different values of this by treating the tax rate τ as a parameter. We can think of three variants of such a policy, and these three will be explicitly used when we come to simulate the MSG2 model:

1. A first variant allows the inbuilt stabilizers to work, operating through the exogenous tax rate, τ.
2. A second variant is a completely rigid-Maastricht rule in which the automatic stabilizers are *entirely switched off*, so that $\tau = 0$ in the simple model. There will still be some endogenous fiscal adjustment in the MSG2 simulations to ensure that the intertemporal budget constraint is not violated.
3. A third variant allows the inbuilt stabilizers to work, but also builds in an activist fiscal rule, representing a fiscal stance which is more strongly reactive to activity than would be the case simply by letting the inbuilt stabilizers work; we can represent this case in this case by a large $\tau > 0$.

Exchange rate determination   The final piece of the model—the BP component—is only of relevance in a floating exchange rate regime, and is shown in equation (3):

$$e = -\theta r. \tag{3}$$

This shows the size of exchange rate appreciation caused by a gap between the home interest rate, r, and the US interest rate (which does not change and so is zero). In a simple Mundell Fleming model with perfect capital mobility and static exchange rate expectations, this effect would be infinitely large, i.e. $\theta \to$ infinity. (It means that, whatever the monetary policy under operation, home interest rates cannot end up different from US interest rates.) But we wish for something more in the spirit of a simple Dornbusch model, in which a gap between home and foreign interest rates is possible and we represent this by a value of θ of less than infinity: we suppose that the interest rate differential is offset by an expectation that the exchange rate will gradually return to its equilibrium level.[3]

A Two-Country European Model

We now present a simple model of a two-country Europe. For simplicity we suppose that the home and foreign countries are of identical size, and have identical parameters. Also, for the present we write the foreign country's LM curve in an identical manner to that for the home country, which includes the possibility of floating, or of fixed exchange rates *with the rest of the world*. Intra-European fixed exchange rate regimes will be described below.

We retain equations (1)–(3) for the home country and add the following equations for the foreign country:

$$y^* = \delta e^* + \sigma(e^* - e) - \gamma r^* + \phi(1-\tau)y^* - \mu y^* + \eta y + v^*, \tag{4}$$

$$r^* = \beta y^*, \tag{5}$$

$$e^* = -\theta r^*. \tag{6}$$

The European Model in Sums-and-Differences Form

By utilizing the simple sums-and-differences approach due to Aoki (1980), we may write the above two-country model in a different way, which will be helpful in what follows. We first add the equations for the home and the foreign country, to obtain a model for Europe as a whole (a "sums" model) and then subtract the equations for foreign from those for home, to obtain a model for the differences between home and foreign (a "differences" model). Again we leave any alterations of the interest rate equation, due to intra-European fixed exchange rate arrangements, until later. We denote sums by the subscript s and differences by the subscript d; thus for any variable x, $x_s = x + x^*$, and $x_d = x - x^*$. The sums model may now be written as

$$y_s = \delta e_s - \gamma r_s + \phi(1-\tau)y_s - \mu y_s + \eta y_s + v_s, \tag{7}$$

$$r_s = \beta y_s, \tag{8}$$

$$e_s = -\theta r_s, \tag{9}$$

and the differences model as

$$y_d = (\delta + 2\sigma)e_d - \gamma r_d + \phi(1-\tau)y_d - \mu y_d - \eta y_d + v_d, \tag{10}$$

$$r_d = \beta y_d, \tag{11}$$

$$e_d = -\theta r_d. \tag{12}$$

There are two important ways in which the structure of the sums model differs from the structure of the earlier basic open-economy model. First here are *positive* income feedback effects. Whereas partner country output is given in the single country model described above, here it is endogenous: in the sums model the foreign interaction effect augments disturbances to the IS curves: demand shocks are "bottled-up" within the European sum. An alternative way of putting this point is that the sums model is less open than the basic open-economy model, and so damping due to imports is less. Second, competitiveness effects are *smaller*. In the sums model—for Europe as a whole—these depend only on European competitiveness, and they are only half as large as in the original model. This is because intra-European competitiveness effects are of no concern for the European total: trade as a proportion of total demand for output is therefore cut.

Similarly, the structure of the differences model differs from the structure of the earlier basic open-economy model in two ways. First income feedback effects due to imports now *dampen* the results of the basic open economy model: because disturbances to output are positively transmitted between the European countries, a rise in output in one country leads to a smaller change in the differences in outputs between countries. Second, competitiveness effects are *stronger*. In the differences model these effects depend not only on the difference between the countries' demands for rest of the world (US) goods (by analogy with the basic model), but also on home country demand for foreign goods, and foreign demand for home goods (and both of these latter effects depend on the difference between the countries' competitivenesses[4]).

3. Simulating Macroeconomic Shocks with the MSG2 Model Under Different Monetary and Fiscal Regimes

The MSG model was developed by Warwick McKibbin and Jeffrey Sachs in two distinct stages. The first model called MSG formed the basis of a number of papers by the authors in the mid-1980s. This model is also the version which participated in the international model comparison project reported in Bryant et al. (1988, 1992). This earlier model was a macroeconomic model of the world economy with rational expectations in the foreign exchange market. The parameters were essentially reduced-form parameters calibrated to the estimates of existing macroeconometric models.

This model was then completely reconstructed beginning in 1986, following the approach taken by CGE modelers which focuses on individual optimization by economic agents. This new model, called MSG2, is reported in McKibbin and Sachs (1991). The MSG2 model, like CGE models, is based more firmly on microfoundations than the standard macroeconometric model. But it is also dynamic: it is described by its authors as a dynamic general equilibrium model of a multiregion world economy. Explicit intertemporal optimization of agents forms the basis of structural behavioral equations: in contrast to static CGE models, time and dynamics are of fundamental importance. In addition, as in all macroeconomic models, there is an explicit treatment of the holding of financial assets including money.

In order to be able to fit macro time series data, the behavior of agents is modified to allow for short-run deviations from optimal behavior either due to myopia or to restrictions on the ability of households and firms to borrow at the risk-free bond rate on government debt. Deviations from intertemporal optimizing behavior take the form of rules of thumb which are consistent with an optimizing agent that does not update predictions based on new information about future events. These rules of thumb are chosen to generate the same steady-state behavior as optimizing agents. Actual behavior is assumed to be a weighted average of the optimizing and rule-of-thumb assumption. For example, aggregate consumption is a weighted average of consumption based on wealth and consumption based on current disposable income. The other key modification to the standard market-clearing assumption in CGE models is the allowance for short-run nominal wage rigidity in different countries. As a result, the model has a mix of Keynesian and classical properties.

The version of the model used in the current paper consists of the following country blocs: the United States, Japan, Germany, the United Kingdom, France, Italy, the rest of the EMS (denoted REMS),[5] Canada, the rest of the OECD (denoted ROECD),[6] nonoil developing countries (denoted LDCs),[7] oil exporting countries (denoted OPEC),[8] and eastern European economies including the Commonwealth of Independent States.[9] The model is of moderate size (about four dozen behavioral equations per country region). The main features of the model are as follows.

(1) Both the demand and the supply features of the main economies are explicitly featured (which is a reason why it is an attractive vehicle for a much fuller treatment of the questions at hand than the simple demand model set out in the previous section).
(2) Demand equations are based on a combination of intertemporal optimizing behavior and liquidity constrained behavior (which means that there are income feedbacks of the kind identified in the IS curve above). There are explicit demand equations for both exports and imports which depend on foreign demand and relative prices in the manner presented above.
(3) Prices adjust rapidly to equate the supply of and demand for both assets and produced goods, but wages adjust sluggishly to imbalance between the supply of and demand for labour. This gives the model something of the flex-output character of our simple fixed-price model above, *but only in the short run.*
(4) The supply side takes explicit account of the effect of exchange rate change on the costs of imports, and thus on prices. It also explicitly considers the intertemporal effects of the accumulation of physical capital and also of the use of intermediate capital goods inputs (whose role in international trade is explicitly modeled).
(5) Major flows such as private physical investment and public physical investment, fiscal deficits, and current account imbalances cumulate into stocks of capital (and equity), infrastructure capital, government debt, and net external debt, and as a result the level and the composition of national wealth changes over time.
(6) Wealth adjustment determines stock equilibrium in the long run but also feeds back into short-run conditions through the effect of asset proportions on prices in forward-looking share markets, bond markets, and foreign exchange markets.
(7) Asset markets are linked globally through the high international mobility of capital.

The careful attention to microfoundations in the model and the empirical basis of the model makes it particularly suited to checking the usefulness and empirical relevance of the insights which we have obtained from our simple model.

4. The Effect of a Demand Shock in a Single Open Economy

A Demand Shock in Our Simple Open-Economy Model

Fixed exchange rates Suppose that this economy belongs to a fixed exchange rate regime. As already explained, this means that in our model we set θ to zero so that the exchange rate does not move, and also set β to zero so that—as required by the fixed exchange rate—the nominal interest rate does not move.

Consider the first column of Table 1. The shock is of course not undamped. Considerations of the entry, with θ and β set to zero, gives a result of $dy/dv_1 = 1/[1 - \phi(1 - \tau) + \mu]$. Stabilization is provided by:

Table 1. *Effects of Demand Shock in Various Models*

Effect on:	Basic open-economy model	Sums model	Differences model
y	$1/\Delta$	$1/\Delta_s$	$1/\Delta_d$
e	$-\theta\beta/\Delta$	$-\theta\beta/\Delta_s$	$-\theta\beta/\Delta_d$

$$\Delta = 1-\phi(1-\tau)+\mu+\beta[\gamma+(\delta+\sigma)\theta]>0.$$
$$\Delta_s = 1-\phi(1-\tau)+\mu-\eta+\beta(\gamma+\delta\theta)>0.$$
$$\Delta_d = 1-\phi(1-\tau)+\mu+\eta+\beta(\gamma+\delta\theta+2\sigma\theta)>0.$$

(a) leakages to imports, which depend on the degree of openness, μ, of the economy; and

(b) leakages to taxes—these arise from the inbuilt stabilizers represented by $\tau>0$; they are cancelled if the fiscal stabilizers are switched off and enhanced in size if these are strengthened.

What is of course missing from this fixed-price account is an essential mechanism of further stabilization under fixed exchange rates—the erosion of competitiveness caused by higher prices, and the feedback of this in reducing demand.

Free floating exchange rates   In the floating exchange rate regime the shock is further damped, provided that the tax rate parameter is the same as in the fixed exchange rate case. This happens as a result both of interest rate increases ($\gamma > 0$) and of exchange rate appreciation ($\theta > 0$). But floating exchange rates do *not* here fully isolate the economy from a demand shock, in contrast to what is predicted in the simple Mundell Fleming model, where, in effect, θ is infinitely large. If θ is very large then the exchange rate will appreciate by enough to choke off any increase in nominal income and here—with fixed prices—that means that output cannot increase.[10]

The extent to which the damping is more in the floating rate case is of course an empirical question. Similarly so is the effect on this comparison of different sizes of the tax parameter τ.

Simulating a UK Consumption Shock with the MSG2 Model Under Different Monetary and Fiscal Regimes

The shock   We consider a shock to UK consumption over four years with the following size as a proportion of GDP over years 1 to 4: 2%, 1.5%, 1.0%, and 0.5%. The size of this shock is chosen with the UK experience of the late 1980s in mind. Its time profile deliberately captures the fact that such shocks are autocorrelated rather than white noise, and a period of four years corresponds to the stylized facts of that UK experience.[11] It is not slowly phased in or suddenly phased out, in order to avoid odd dynamic effects in a forward-looking model.

The UK policy regimes   Under the floating exchange rate regime we suppose that the UK monetary authorities adhere to the following interest rate rule: $r = \beta(p+y)$, where p is the GDP deflator (measured in proportional deviations from an initial position), and the parameter β is set at the value 1.66. This rule implies that the interest rate rises whenever the demand for money (which depends on nominal income) rises relative to a fixed supply.[1] An alternative interpretation of this rule is that it is an explicit nominal income rule, in which the nominal interest rate is raised whenever nominal income is above target (Taylor, 1994).[2]

We assume, under the floating exchange rate regime, that the fiscal authorities allow the inbuilt stabilizers to operate.

Under the fixed exchange rate regime we assume that the UK exchange rate is pegged to the deutschmark, with the same interest rate as Germany,[12] and that the UK fiscal authorities operate one of three different regimes:

(1) In the first regime, known as Fixed1, the inbuilt stabilizers are allowed to work, so that the value of τ is merely that implicitly given in the MSG2 model.
(2) In the second regime, known as Fixed2, the automatic stabilizers are turned off, effective setting τ to zero.[13]
(3) The third regime, known as Fixed3, is illustrative of a more activist fiscal policy rule. It allows the inbuilt stabilizers to work, but also builds in a rule which allows fiscal stance to be more strongly reactive to activity than would be the case simply by letting the inbuilt stabilizers work. This adds to the value of τ an increase in taxes equal to *half* of any increase in GDP, over and above the operation of the inbuilt stabilizers.

The shock to be simulated is supposed to happen to the UK alone. But in simulating this shock on the MSG2 model we need to make some assumption about what is happening in Europe and in the rest of the world. In the case of the UK operating the float regime, it is assumed that all other countries operate an identical float regime. In the case of the UK operating a fixed exchange rate regime we suppose that the UK pegs to Germany, but that all other countries (not only including Germany but also including all other European countries) continue to operate the exactly the same floating regime.[14]

Results   Consider the results for the floating regime shown in Figure 1.[15] This shows that the effect of the consumption shock is to cause an immediate rise in GDP of 1% (equal to exactly half of the initial consumption shock). There is an immediate appreciation in the nominal exchange rate of more than 6%, which speedily disappears as the shock dies away, and has a small negative echo. Prices rise (even although actually import prices fall following the large appreciation).[16] The trade balance bears the brunt of the cushioning of the economy from the shock, worsening by nearly as much as the shock itself. Because the effects of real exchange rate change on the trade balance are in reality lagged, there is a negative echo of the isolation effect in causing the change in real GDP to go negative after the third year.

Next consider the results for the fixed exchange rates. In variant Fixed1 the effect of the consumption shock is to cause an immediate rise in GDP of approximately $2\frac{1}{2}\%$, about two and a half times the size of the change in the floating rate regime. This is even after the dampening effects of the inbuilt fiscal stabilizers, and of competitiveness, work through.[17]

Now consider variant Fixed2, in which the inbuilt stabilizers are switched off. This shows how important these are, for a single open economy, in effecting the size of all of the fluctuations—to GDP, to net exports, and to prices. All fluctuations are magnified in variant Fixed2 by a factor of about one-third. Or to put the matter another way, without the inbuilt stabilizers the fluctuations are about three and a half times

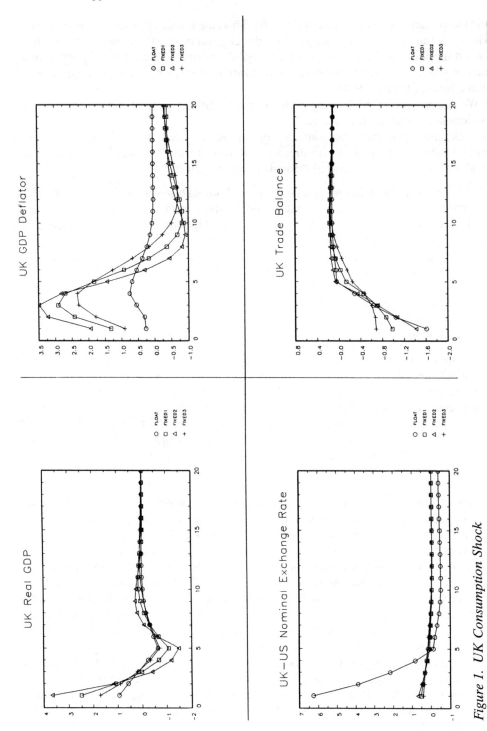

Figure 1. UK Consumption Shock

Key: All plots show percent deviation from base except that for the Trade Balance which shows deviation from base as a percent of GDP. For explanation of regimes Float, Fixed1, Fixed2 and Fixed3, see the text on page 63

as large as those under flexible exchange rates rather than only two and a half times as large.

Variant Fixed3 shows how hard it is to further damp the effects by further fiscal activism. The fluctuations are damped compared with variant Fixed1 by a factor of only about one-third; even with this extremely large degree of fiscal activism they remain much larger than the fluctuations under floating rates.

5. The Effects of Demand Shocks in Europe

We now turn to extend our insights from the single open economy to consider the operation of different monetary and fiscal regimes in a multicountry Europe, using both our simple theoretical model and the MSG2 model.

Three European Regimes in Our Simple Theoretical Model: Floating Exchange Rates, EMU, and Hard ERM

We will consider three regimes. In our simple theoretical model, extended to a two-country Europe, they operate as follows.

(1) In the floating exchange rate regime, there is free floating everywhere. Thus in this regime the central banks in both European countries follow an interest rate rule of the kind displayed in equation (2). This means that, *in both sums model and the differences model*, monetary policy is conducted exactly according to this equation. In this regime the built in stabilizers operate, so that in this regime $\tau > 0$ for that reason.

(2) In the EMU regime, interest rates are the same in both countries. They are set by a European central bank. This bank follows an interest rate rule of the kind displayed in equation (2), but with one crucial amendment. Since output is measured as percentage deviations from base, the analogs to equations (2) and (5) for the home and foreign country become

$$r = \beta\left(y + y^*\right)/2, \tag{13}$$

$$r^* = \beta\left(y + y^*\right)/2. \tag{14}$$

This rule governs European monetary policy as a whole, and thus appears only in the sums model; giving

$$r_s = \beta y_s. \tag{15}$$

In the differences model, by contrast

$$r_d = 0. \tag{16}$$

European exchange rates are locked together so that $e = e^*$, and therefore, in the differences model, $e_d = 0$. We thus have

$$e_s = -\theta r_s, \tag{17}$$

$$e_d = 0. \tag{18}$$

In this EMU there is also an agreed Europe-wide fiscal policy rule. We consider: (a) the possibility in which the inbuilt stabilizers are allowed to work, so that in this variant τ takes the same value as it does in the floating rate model, *in both the sums and differences models*; (b) the possibility that the inbuilt stabilizers switched off in

both the sums and the differences models; and (c) the possibility that the inbuilt stabilizers are enhanced.
(3) In the "hard" ERM regime, European exchange rates are again locked together so that $e = e^*$ and $e_d = 0$. In this union, interest rates are the same in both countries, but they are now set by the home country according to equation (2) for that country alone. This rule thus governs European monetary policy as a whole, and thus drives interest rates in the sums model.[18] The LM–BP parts of the ERM model thus consist of equation (2) and equations (16)–(18). For this ERM we also consider the possibility of τ taking different values.

Sums Shocks, Differences Shocks, and the Design of EMU

We may now use our simple two-country model to examine the effects of shocks which are perfectly positively correlated between European countries (sums shocks) and shocks which are perfectly negatively correlated between European countries (differences shocks).

Sums shocks   Results for sums shocks are shown in the second column of Table 1. We see that, for whatever regime, the multiplier $1/\Delta_s$ is larger than that for the basic open economy, both because the intercountry demand effects, captured by the η term, are "bottled up" within Europe, and also because the effects of currency appreciation on demand are damped: they work only on trade beyond Europe and not on trade between the countries. This suggests that a floating regime, for Europe as a whole, has the advantage of dampening Europe-wide shocks (even although the exchange rate movements, per unit shock, would be larger than for the basic open economy).

Differences shocks   Results for differences shocks are shown in the third column of Table 1. We see that, for whatever regime, the multiplier $1/\Delta_d$ is smaller larger than that for the basic open economy, both because the intercountry demand effects, captured by the η term, are weakened within Europe (as output movement in one country drags up output in the other), and also because the effects of currency appreciation on demand are strengthened: for each country they work on trade within Europe as well as on trade beyond. This suggests that the implications of asymmetric shocks for output fluctuations under fixing exchange rates within a (tightly integrated) Europe would be less serious than would be imagined simply by analyzing the structure of the shocks and imagining them impinging on a single open economy.

Implications for the design of EMU   Our simple model enables us to describe the macroeconomic policy properties of EMU in the following simple way. It has two components. The *stronger* shock-absorbing properties of floating exchange rates are deployed against the effects of demand shocks that are correlated between the countries, shocks which are anyway stronger because of the inbuilt intercountry spillovers which magnify them. By contrast, merely the *weaker* shock-absorbing properties of fiscal feedbacks operating on their own are deployed against negatively correlated demand shocks, which are already weaker because the inbuilt intercountry spillovers dampen them. But having switched off intercountry monetary stabilization, it is important that fiscal shock absorbing is not also weakened, or in the limit switched off, by constraints and rules on the allowed degree imposed by the rules of the monetary union itself.

A Home-Economy Demand Shock in Our Simple European Model

In a European monetary union, the absence of a separate monetary policy ideally suits the response by the common monetary authorities to shocks which are common to all countries. It prevents the authorities in the different countries from responding individually to separate shocks; response by the central monetary authorities to European demand disturbances resulting from demand shocks in one of the economies may impose costs on the other economy. One way of examining the seriousness of this is to examine the conditions under which a positive shock in one country will give rise to a negative shock in the other country. We would argue that, under these circumstances, there would be at least a *prima facie* case for concern.

We therefore use our simple model to examine whether a demand shock in the home economy is negatively transmitted to the foreign economy, under various fiscal and monetary regimes. Results are displayed for the three regimes in Table 2. A demand shock in Germany alone, v, might be thought as composed of positive $v_s = v/2$ in the sums model and a simultaneous positive $v_d = v/2$ in the differences model. Use of this fact enables us to get some insight into the results from the simpler outcomes in Table 1.

Floating The outcomes in Table 2 show that a home-country demand shock causes a rise in home-country output and an appreciation of the home country's currency. It also causes a rise in foreign-country output and an appreciation of the foreign exchange rate.

These results may be immediately understood by considering the sums and differences results in Table 1. In the sums model, the foreign demand effects bottle-up and magnify the shock; and the currency appreciation caused by the interest rate rise has a dampened effect on demand (compared with the simple open-economy model) because there are no intercountry effects to activate. By contrast, in the differences model the income feedbacks are dampened and the exchange rate demand-stabilizing effects are enhanced, because of the between-country competitiveness effects. There-

Table 2. Effects of Demand Shock under Various Regimes

Effect on:	Floating	EMU	ERM
y	$\dfrac{\Delta}{\Delta_s \Delta_d}$	$\dfrac{1-\phi(1-\tau)+\mu+\left[(\gamma+\delta\theta)\beta\right]/2}{\Delta_s \Delta_d^{emu}}$	$\dfrac{1-\phi(1-\tau)+\mu}{(\Delta-\beta\sigma\theta)\left[1-\phi(1-\tau)+\mu\right]-\eta\left[\eta-(\gamma+\delta\theta)\beta\right]}$
e	$\dfrac{-\theta\beta\Delta}{\Delta_s \Delta_d}$	$\dfrac{-\theta\beta\left\{1-\phi(1-\tau)+\mu+\left[(\gamma+\delta\theta)\beta\right]/2\right\}}{\Delta_s \Delta_d^{emu}}$	$\dfrac{-\theta\beta\left[1-\phi(1-\tau)+\mu\right]}{(\Delta-\beta\sigma\theta)\left[1-\phi(1-\tau)+\mu\right]-\eta\left[\eta-(\gamma+\delta\theta)\beta\right]}$
y^*	$\dfrac{\eta+\beta\sigma\theta}{\Delta_s \Delta_d}$	$\dfrac{\eta-\left[\beta(\gamma+\delta\theta)\right]/2}{\Delta_s \Delta_d^{emu}}$	$\dfrac{\eta-(\gamma+\delta\theta)\beta}{(\Delta-\beta\sigma\theta)\left[1-\phi(1-\tau)+\mu\right]-\eta\left[\eta-(\gamma+\delta\theta)\beta\right]}$
e^*	$\dfrac{-\theta\beta(\eta+\beta\sigma\theta)}{\Delta_s \Delta_d}$	Same as for e above	Same as for e above

$\Delta_d^{emu} = 1-\phi(1-\tau)+\mu+\eta > 0.$

Δ, Δ_s, and Δ_d are defined under Table 1.

We assume in the text that $(\Delta-\beta\sigma\theta)\left[1-\phi(1-\tau)+\mu\right]-\eta\left[\eta-(\gamma+\delta\theta)\beta\right] > 0.$

fore, subtracting the differences results from the sums results, a positive output effect remains for the foreign European country. Put very simply, the rise in home-country output, and the relative appreciation of the home-country currency, both increase the demand for foreign-country goods.

EMU   Notice that the behavior of the sums model is *completely independent* of whether there is floating, or whether there is EMU. The effects of intra-European monetary arrangements on aggregate European behavior completely wash out (subject to the proviso that the inbuilt fiscal stabilizers operate in the same way under the two cases which is why the comparison is with the EMU1 variant). In the differences model there are no exchange rate feedbacks.

There are two differences between the sums model results and the differences model results under EMU. Monetary dampening exists in the sums model, but not, because of EMU, in the differences model. This will tend to drag down output in the foreign country. But the income feedbacks in the differences model tend to pull the foreign country's output up with the home country's output. If the former effect dominates, then y^* will fall. Simple calculation reveals that this is so if

$$\beta(\gamma + \delta\theta) > 2\eta. \tag{19}$$

It is apparent that the degree of fiscal stabilization, governed by the size of τ, does not affect whether this condition holds. But, as we shall see below, the important point about fiscal stabilization is that, when this condition fails, it damps the resulting fall in foreign output, y^*.

ERM   It is intuitively clear that foreign output is more likely to fall in the ERM regime than the EMU regime. In more detail the reason is that the effects of intra-European monetary arrangements on overall European interest rates no longer wash out. The more does home income rise relative to foreign income, the higher is the rise in home interest rates and the larger the fall in foreign income. Of course higher home income directly increases home demand for foreign goods, but it also raises home interest rates more strongly than in the EMU regime. These arguments are confirmed by column three of Table 2, which shows that y^* falls if

$$\beta(\gamma + \delta\theta) > \eta, \tag{20}$$

which is a condition twice as strong as that in equation (19).

Simulating the GEMU Shock Under Different European Monetary and Fiscal Regimes

The GEMU shock   This is a big shock. But in portraying the macroeconomic operation of Europe under different regimes we are using as the shock a representation of something that has actually happened, as a strong test of the kinds of strains that could be imposed.

There are two elements to the shock. First a temporary hump, rising quickly to a maximum of 6% of GDP and then falling over a four-year period in the transition to a new long-run equilibrium (representing the long-term effect of unification). After year 10 there is a gradual fiscal tightening, and in addition, there is a further rise in taxes—not shown—to cover the extra debt interest incurred, so that, by year 20, steady-state budget balance has been approximately regained. The total area under the two lines shown over the 20 years, which shows the approximate size of the shock assumed, is about 45% of one year's GDP.

We will display the outcomes for Germany (the "home" economy) and for the UK, a representative foreign European economy.

The European policy regimes Under floating we suppose that all European monetary authorities adopt the interest rate rule $r = \beta (p + y)$, where $\beta = 1.66$, and that the fiscal authorities in all countries allow the inbuilt stabilizers to operate. Under EMU we assume that European monetary policy is unified and that there is a common currency. The European Central Bank authorities operate the above rule, but now the $(p + y)$ term is the average for Europe as a whole. We investigate outcomes when all of the European fiscal authorities together operate one of three different fiscal variants already described.

(1) In the first variant, EMU1, the inbuilt stabilizers are allowed to work, so that $\tau > 0$. These could be operated by national governments or by a central fiscal authority in Brussels operating through the exogenous tax rate.
(2) In the second variant, EMU2, the automatic stabilizers are turned off, effectively setting τ to zero. This is designed to capture the operation of an admittedly extreme form of fiscal noncoordination within a monetary union—with no central fiscal federalism allowed and also with no agreement obtained for the decentralized operation of automatic stabilizers.
(3) The third variant, EMU3, is illustrative of a more activist fiscal policy rule. It allows the inbuilt stabilizers to work, but also builds in a rule which allows fiscal stance to be more strongly reactive to activity than would be the case simply by letting the inbuilt stabilizers work. This adds to taxes one half of any rise in real GDP, over and above the operation of the inbuilt stabilizers.

Under ERM we assume that European monetary policy is unified, and that there is a common currency. The European Central Bank authorities operate the above rule, but now the $(p + y)$ term is that for Germany alone. We investigate the outcomes for three different regimes ERM1, ERM2, ERM3. The differences between these three regimes are identical to the differences between the regimes EMU1, EMU2, and EMU3.

Results—Floating Consider the results for the floating regime shown for Germany (in Figure 2) and for the UK (in Figure 3). These show that the effect of the reunification shock is almost completely removed by floating. After a pause, there is a rise in GDP of up to 2%, which begins falling back after three years, even although the shock last for much longer. There is an immediate rise in the nominal exchange rate of approximately 15%, which only slowly disappears. German prices actually fall, because import prices fall following the large appreciation. The trade balance bears the brunt of the cushioning of the economy from the shock, worsening by nearly as much as the shock itself.

For the UK, output falls slightly, as compared with the predications of our simple theoretical model which show an increase in foreign output. This suggests that our simple model has missed something: the supply-side spillovers of the UK depreciation relative to Germany in causing a further damping of UK output.

Results—EMU Next consider the EMU results which are shown in the figures. Consider first variant EMU1. The effect of the demand shock is now to cause an immediate rise in GDP of approximately only 3%, only half as much again as under floating. (This compares with a ratio 2.5 times in the difference between fixed and floating in our analysis of the UK consumption shock.) The reason that a move from float to EMU is

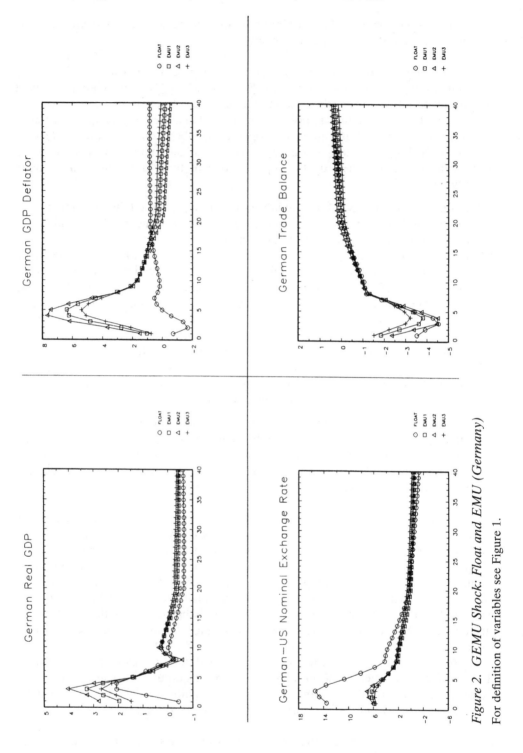

Figure 2. GEMU Shock: Float and EMU (Germany)
For definition of variables see Figure 1.

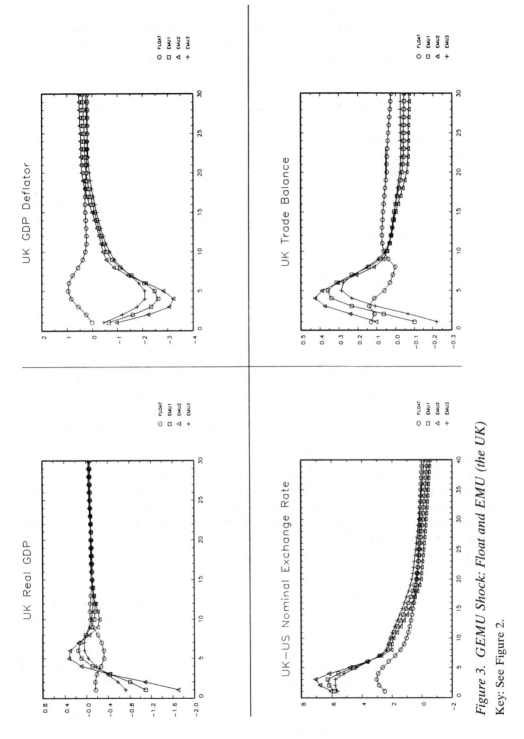

Figure 3. GEMU Shock: Float and EMU (the UK)
Key: See Figure 2.

very different from a move from float to fixed exchange rates is as follows: in EMU, Europe-wide interest rates rise (so that German real interest rates increase nearly as much as in the floating case) and the Europe-wide exchange rate appreciates (more than a third as much as the German exchange rate appreciates in the floating case), hence strongly damping the rise in German output.

For the UK, output now falls significantly, suggesting that income effects are not sufficient to outweigh the effects of exchange rate appreciation, so that the analog of condition (19) does not hold. But relative price movements quickly damp the movements, causing a redistribution of the shock through the real exchange rate, even under fixed nominal exchange rates.

Now consider variant EMU2, in which the inbuilt stabilizers are switched off. German results are not very different from variant 1. But output falls much more strongly in the UK. The fiscal stabilizers are being prevented in redistributing the shock from Germany to the UK, showing empirically the importance of the issue referred to theoretically earlier on. Variant EMU3 shows how, if possible, more active fiscal stabilization could further redistribute the shock.

Results—ERM   Finally consider the results for the "hard" ERM regimes which are shown in Figures 4 and 5. In the hard ERM regime all European countries tie their exchange rates tightly to the DM and German monetary policy now concerns itself with nominal income in Germany alone.

Consider first variant ERM1. The effect of the German shock is now to cause German outcomes which are very similar to those under floating. Furthermore, when we consider the different ERM variants, with different fiscal damping, these also differ little from each other, or from the outcome under floating. The most outstanding thing is the similarity of outcomes. In particular, the whole of the ERM experiences the same size of exchange rate rise as that which Germany would experience if it were to float.

By contrast, under variant ERM1 output now falls considerably in the UK, whereas under floating it hardly fell at all: UK output falls as a proportion of GDP by an amount nearly equal to half of the size of the shock in Germany, as a percentage of GDP. After three or four years UK output has recovered, mainly as a result of the fall in UK prices relative to German prices and the transfer of the shock to the UK through the trade balance. And the less the degree of fiscal stabilization the more does output fall in the UK when the stabilizers are switched off and output in the UK falls massively.

This apparent radical difference in outcomes may be explained as follows, using ideas from our simple model. When Germany floats it obtains two stabilizing benefits from its appreciation. First there is a fall in German competitiveness *vis-à-vis* the rest the world, and second there is a fall in German competitiveness *vis-à-vis* the rest of Europe; both of these effects choke off net exports. Suppose that the only difference between floating and ERM is that the second choking-off did not happen (since other European exchange rates are tied to the deutschmark in the ERM). Then the appreciation of *all of Europe* against the rest of the world would have to be *larger* than the appreciation of Germany on its own when it floats. However, of course, the ERM regime imposes output losses on the rest of Europe. These cause a drop in the demand for German goods, attenuating the necessary rise in the German interest rate, and thereby reducing the required European appreciation, relative to this initial supposition. What happens is that, on balance, the German appreciation in the two cases is the same. But whereas in the first case this appreciation is accompanied by a reduction in the demand for German goods as Germany appreciates against the other countries, in

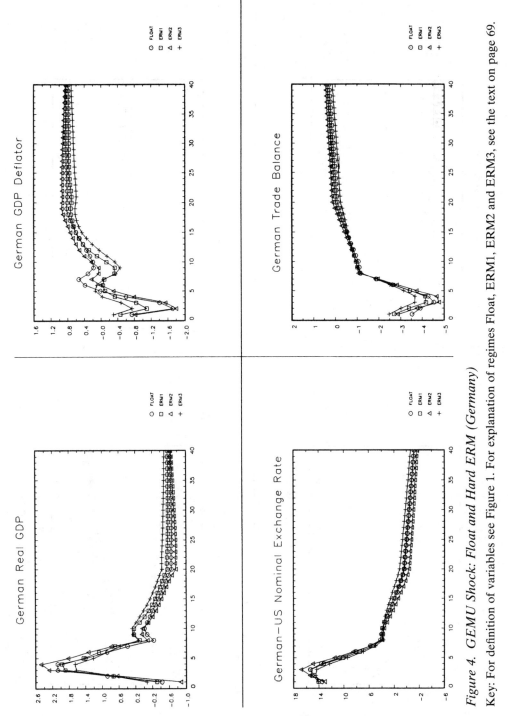

Figure 4. GEMU Shock: Float and Hard ERM (Germany)
Key: For definition of variables see Figure 1. For explanation of regimes Float, ERM1, ERM2 and ERM3, see the text on page 69.

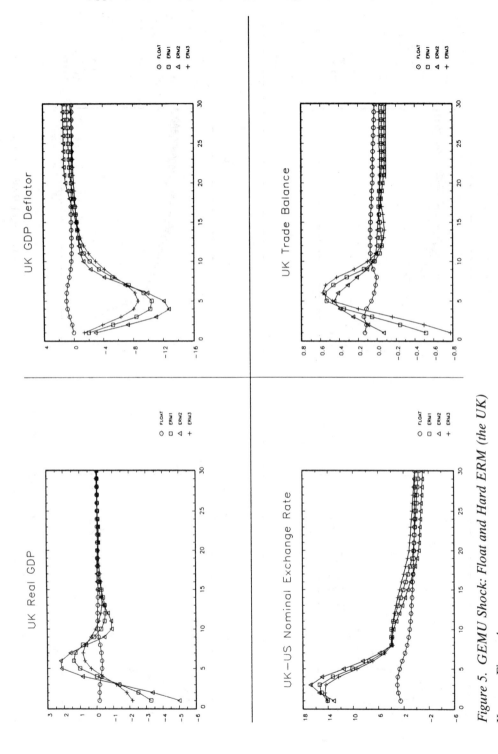

Figure 5. GEMU Shock: Float and Hard ERM (the UK)
Key: see Figure 4.

the second case it is accompanied by a reduction in the demand for German goods caused by the fall in output elsewhere in Europe.[19]

These results provide radical confirmation of the costs which are imposed on other countries if they are constrained, by monetary union, from altering their interest rates relative to the European country experiencing the shock, and at the same time (a) Europe-wide interest rates are raised with reference to the stabilization needs of the countries experiencing the shock, and (b) political constraints are put on the operation of the inbuilt fiscal stabilizers.

6. Conclusion

This paper has examined alternative macroeconomic stabilization rules for demand shocks, for an integrated European region. We have shown that, in a two-country region, positively correlated demand shocks are "bottled up" and magnified, augmenting each other through positive demand spillovers between the countries. Conversely, negatively correlated shocks are dissipated and damped, because positive demand spillovers between the countries tend to dampen regional differences. We have described the operation of EMU as using the relatively strong shock-absorbing properties of floating exchange rates to stabilize the effects of the demand shocks which are correlated between the countries, shocks which are anyway stronger because of the inbuilt intercountry spillovers which magnify them. At the same time it uses the weaker shock absorbing properties of fiscal feedbacks against negatively correlated demand shocks, which are already weaker because the inbuilt intercountry spillovers dampen them. These could be operated by national governments or by a central fiscal authority in Brussels.

We identified two design-risks to such an EMU system.

(1) The first fear is that system-design problems on the fiscal side will lead to the weakening of the fiscal feedbacks operating within each country.
(2) The second fear is that system design problems on the monetary side might leave monetary restraint for Europe as a whole responding to demand disturbance in just one region. We investigated the conditions required such that a positive demand shock in one country should not have a negative, or strongly negative, effect on output in the other country.

Our results suggest that an EMU could be made to function with the properties described above, but that disturbances in one country would have a strongly negative effect on the other country if either of these risks were not dealt with.

Simulations with the MSG2 model bear out the significance of these risks. The show that a demand shock, like GEMU, even if smaller, can have strongly negative effects on output in other European countries if either interest rates are raised to counter the demand shock in the originating country, or if, for some reason, fiscal stabilization is not allowed to be as strong as the inbuilt fiscal stabilizers.

References

Aoki, M., *Dynamic Analysis of Open Economies*, New York: Academic Press, 1980.

Bayoumi, T. and B. Eichengreen, "Shocking Aspects of European Monetary Unification," in Francisco Torres and Francesco Giavazzi (eds.), *Adjustment and Growth in the European Monetary Union*, Cambridge: Cambridge University Press, 1993.

Bryant, Ralph, Dale Henderson, Gerald Holtham, Peter Hooper, and Steven Symansky (editors), *Empirical Macroeconomics for Interdependent Economies*, Washington, DC: Brookings Institution, 1988.

Bryant, Ralph, Peter Hooper and Catherine Mann (editors), *Evaluating Policy Regimes: New Research in Empirical Macroeconomics*, Washington, DC: Brookings Institution, 1992.
Kenen, Peter B., *European Monetary Union After Maastricht*, New York: Group of Thirty, 1992.
McKibbin, Warwick and Jeffrey Sachs, *Global Linkages*, Washington, DC: Brookings Institution, 1991.
Taylor, John, *Macroeconomic Policy in a World Economy*, New York: Norton, 1994.

Notes

1. The coefficient of response used below, 1.66, is equal to the inverse of the interest rate semielasticity in the demand for money, which is 0.6.
2. The coefficient which Taylor uses, around 1.5, is almost exactly equal to that which we have taken from inverting the demand for money in the MSG2 model.
3. In our setup in which all shocks are temporary, this equilibrium will be equal to the old value. Of course in a rational-expectations version of the Dornbusch model, the rational value of θ depends on the speed of adjustment. That is a property of the MSG2 model used below.
4. This is why there is a 2σ term in the IS equation.
5. This group consists of Belgium, Denmark, Ireland, and Luxembourg.
6. This group of countries consists of Australia, Austria, Finland, Iceland, Norway, Spain, Sweden, Switzerland, and New Zealand.
7. Nonoil developing countries are based on the grouping in the IMF *Direction of Trade Statistics*.
8. Oil exporting countries are based on the grouping in the IMF *Direction of Trade Statistics*.
9. These countries are Bulgaria, Czechoslovakia, eastern Germany, Hungary, Poland, Romania, Yugoslavia, and the USSR.
10. What this account misses is that, in a model explicitly identifying price adjustment, choking off an increase in nominal income does not involve choking off a rise in activity, since the resulting currency appreciation will cause a fall in prices, enabling a rise in output.
11. See also to the results of Taylor (1994), and Bayoumi and Eichengreen (1993) for a discussion of the autocorrelation of demand shocks.
12. For a discussion of this assumption, see the next section.
13. It should be remembered that there is still a fiscal response working in the MSG2 model to ensure the intertemporal government budget constraint is satisfied.
14. We could have assumed that the UK pegs to a basket of European currencies, to neutralize any consequences of a fixed-exchange-rate UK dragging Germany in a different direction from the rest of Europe. It would probably have been best of all to simulate with the UK pegged to the US.
15. Further results are available from the authors on request.
16. It is true that the appreciation is large enough to cause a fall in consumer prices, but what matters for the operation of the interest rate rule which is driving the economy is the movement of the output price deflator, and in this the effects of higher output dominate.
17. The latter process works as follows. There is immediate inflation, which causes a gradual increase in the price level which is up by 3% by year 3 and still up 2% by year 5, by which time the consumption shock has already completely died out. Net exports now do some cushioning of the economy from the shock, but not more than half of what happened before. In addition, the lag in the price level responses, coupled with the fact that the effects of the real exchange rate change on the trade balance are lagged, causes a pronounced antishock to GDP. The GDP goes below baseline after the third year, even although the shock remains until the fourth year.
18. This makes it impossible to solve the sums and differences models separately, meaning that the calculations leading to column 3 in Table 2 are rather tedious.
19. Note that the results shown in the figures are influenced by the fact that the UK relies on US trade more than does Germany, so that the appreciation of sterling leads to a disproportionately large reduction in output.

Review of International Economics, Special Supplement, 77–91, 1997

European Nominal and Real Convergence: Joint Process or Rival Dynamics?

*Martine Carré**

Abstract

Recent empirical studies have revealed that the convergence speeds of nominal and real variables are fairly different. This paper studies the temporal evolution of the mutual influence between the convergences of a nominal and of a real variable. It refers first to σ-convergence analysis. In order to compare nominal and real paths, the evolutions of the cross-sectional variances of the two variables are connected. Then, nominal and real convergence processes are studied, as reciprocally conditioned within the same system. Nominal and real convergence were negatively correlated in the 1980s, while the 1990s have been characterized by a simultaneous convergence movement.

1. Introduction

Countries wishing to join the Economic and Monetary Union (EMU) before the end of the 1990s have to satisfy a certain number of convergence criteria specified in the Maastricht Treaty. These criteria were elaborated in order for potential members to face a monetary union with similar structures, compatible with a common economic policy. Indeed, a common response to exogenous shocks could be inappropriate if the members are not close enough. In this view, convergence among European economies is a necessary stage before the final phase of a monetary union. The Treaty underlines implicitly that exchange rate and price stability will favor growth and economic integration. Under this assumption, nominal convergence will lead to real convergence among European countries. The opposite, however, cannot be excluded *a priori*.

A large number of empirical studies have attempted to assess whether European countries have converged. Some have used the cointegration approach (Ardeni, 1992); others have estimated models with variable parameters (Haldane and Hall, 1991; Hall et al., 1992). These studies, whether using time series or cross-sectional data, generally point to the existence of nominal convergence together with real divergence during the 1980s. A common feature of all these studies is that they analyzed these two convergence processes independently. An exception, not directly related to the issue of European integration, is Barro (1995). He finds that poor countries grow faster than rich countries, and tend thereby to catch up, but only if they have favorable settings for government policies. He shows empirically that although the adverse influence of inflation on growth looks small, the long-term effects on standards of living are substantial.[1] Finally, he is concerned exclusively with the effect of inflation on growth but not the reverse. A contribution of this paper is to consider the relative evolution of nominal and real variables, in terms of conditional convergence. I propose a simulta-

---

* Carré: MAD, Université Paris-1, 90 rue de Tolbiac, 75634 Paris Cedex 13, France. Tel: (33)-(01)-40771884; Fax: (33)-(1)-45847889; Email: carrem@univ-paris1.fr. I wish to thank P. Y. Hénin for helpful comments and suggestions. Moreover, I benefited from discussions with F. Collard, J. F. Jacques, and J. M. Tallon. I thank participants at the MAD seminar, the AFSE Conference on European Integration (Nantes, June 1995) and the T2M Conference in Paris (January 1996). I am most grateful to two anonymous referees for their constructive comments that have led to significant improvements in the paper.

neous examination of the convergence process for nominal and real variables in a bivariate system. The study of a joint system enables us to evaluate empirically the link between nominal and real convergence.

The existence of such a link can have several theoretical justifications. As Blanchard and Muet (1993) argue, if and when European countries become part of a common currency area, "competitive disinflation" will be the only mechanism left to improve competitiveness. The convergence necessity expressed by the Treaty should lead European countries to give up expansionary and inflationary policies. But a country which fixes its exchange rate, starting with higher inflation than its trading partners, will initially lose competitiveness further and experience an increase in unemployment. Indeed, while inflation will decline, it will for some time remain higher than foreign inflation, leading to further real appreciation. Eventually, however, inflation will become less than foreign inflation, competitiveness will improve and demand will increase. Strategies of this type tend to be slow to give results. Furthermore, they seem to be more successful at bringing down inflation than at favoring economic growth. The one clearest gain from "competitiveness through disinflation" has been disinflation.

Indeed, there is a danger that the disinflationary adjustment process could lead to short-term unemployment becoming persistent. In the long run, one can fear the appearance of high-unemployment zones, in which the long-term unemployed make up a very low level of human capital. The long period of low activity might depress investment, and slow down capital accumulation for a long time. This could affect the long-term growth of countries that were inflationary at the beginning. Cyclical GNP depression could have persistent and structural effects. Feldstein (1992) among others notes that countries with the worst initial conditions could well be driven to low growth paths with high unemployment. An increased dispersion of unemployment rates will, *ceteris paribus*, hinder the effect of concerted policies, and could prevent countries with the highest rates from eventually joining the monetary union (Heylen et al., 1995).

Could the deflationary constraint imposed on member countries constitute a threat to the process of real convergence during the transitory phase? It is important to stress that, if nominal and real convergence are negatively correlated, nominal constraints imposed by Maastricht could amplify real divergences. In terms of economic policy, the issue is to choose between a policy that would aim at nominal convergence, and a policy that would directly promote the convergence of standards of living and activity levels; i.e., that would directly seek real convergence.

The aim of this paper is not to test a particular theory, but to assess the existence of a link between nominal and real convergence. It focuses on the temporal evolution of the mutual influence between nominal and real dynamics, and estimates a bivariate system including a real variable and a nominal one. I chose, as the real variable, the logarithm of per capita GNP. The nominal variable is the inflation rate.[2] Analysis of this system reveals the important influence of the initial per capita GNP level in the nominal convergence equation for most of the period.

I first define the method used in this paper to test nominal and real convergence. I then evaluate empirically the paths followed by the different processes, using the σ-convergence methodology. In order to compare nominal and real paths, the cross-sectional variances evolutions of the two variables are connected in a common figure. This sets the scene for the last section, in which I estimate a system including both nominal and real variables using a two-equation model, which generalizes Barro's equation for β-convergence. The two processes (real and nominal) are then seen as joint processes in a unique bivariate system. This allows an estimation of the mutual influence between the two types of movements. Their evolution over time is then analyzed.

2. Interpreting Convergence Tests

Defining and Testing Convergence

Convergence has been defined in two different ways (Bernard and Durlauf, 1996). The first considers convergence as "catching up," whereas the second definition sees it as the equality of long-run forecasts for countries at a fixed time.

In the sense of the first definition, countries i and j converge between dates t and $t + T$ if disparity at date t of a variable (denoted $y_{i,t}$) is expected to decrease. In short, if $y_{i,t} > y_{j,t}$, then there is convergence between t and $t + T$ if

$$E\left(y_{i,t+T} - y_{j,t+T}\middle|I_t\right) < y_{i,t} - y_{j,t},$$

where I_t denotes all information available at t.

Countries i and j converge in the sense of the second definition if

$$\lim_{k \to \infty} E\left(y_{i,t+k} - y_{j,t+k}\middle|I_t\right) = 0.$$

Cross-sectional methodologies are associated with the first definition. They consider the behavior of the difference of a given variable, say GNP, between two countries, over a fixed time interval, and equate convergence with the tendency of this difference to narrow. In this regard, the σ-convergence à la Barro (1991) is certainly the most intuitive approach. It requires a decrease over time of cross-dispersions for some given variables,[3] so, if we let σ_t be the empirical cross-variance of a variable, we can observe a σ-convergence movement when the ratio σ_{t+j}/σ_t is less than 1, with $j > 0$. A second type of cross-sectional methodology considers the traditional β-convergence equation, coming from the neoclassical growth model. The focus is then the link between an economy's average growth and its initial income, through the β-convergence equation as Barro (1991) defined it:

$$T^{-1}\left(y_{j,T} - y_{j,0}\right) = a + by_{j,0} + u_{j,T}, \tag{1}$$

which tests the convergence of a variable, noted y, and where u is white noise. β-convergence occurs when there exists a significant negative correlation between the initial level of a variable and its average rate of increase. Thus, the weaker the initial level of a variable, the higher is its average rate of increase.

The concepts of σ- and β-convergence are clearly different. Quah (1993) showed how it is possible for a negative relationship between initial income and growth to be compatible with a stable cross-sectional variance in output levels. We can note, with Bernard and Durlauf (1996), that intuitively this can hold because shocks to country-specific growth rates can offset the effect of this negative coefficient. This can easily be proved, using the formulation of Lichtenberg (1994), by looking at the traditional β-convergence cross-sectional regression:

$$y_{i,t+1} = \left(1 + \beta\right)y_{i,t} + u_i, \tag{2}$$

where u is white noise, and σ_u^2 its variance. Let σ_{t+1} (resp. σ_t) be the standard deviation of y_{t+1} (resp. y_t). One has $\sigma_{t+1}^2 = (1 + \beta)^2\sigma_t^2 + \sigma_u^2$, and therefore $\sigma_t^2/\sigma_{t+1}^2 = [1 - (\sigma_u^2/\sigma_{t+1}^2)]/(1 + \beta)^2$.

The σ-convergence depends not only on the sign of β but also on the variance of the noise relative to the variance of the estimated variable. Thus, β-convergence is meant to capture the idea of a long-term convergence tendency, which can be disturbed momentarily by exogenous shocks. The σ-convergence analysis takes these shocks into account. This fact can be held against the σ-convergence concept, being too sensitive to short-term shocks and thus hiding long-term links.[4]

Moreover, using a dynamic version of Galton's fallacy, Quah (1993) and Hart (1995) establish that a negative initial level coefficient in β-convergence regressions is consistent with an unchanging cross-sectional distribution. Starting from equation (2), where we let $\beta' = 1 + \beta$, we observe[5] that $\sigma_{t+1}^2/\sigma_t^2 = \beta'^2/\rho^2$, where ρ is the correlation coefficient between $y_{i,t+1}$ and $y_{i,t}$.

The change in the variance of the variable can then be decomposed into a regression effect—illustrated by the coefficient β' in the regression—and a size mobility effect illustrated by the correlation coefficient ρ. It is possible for countries to diverge if $\beta' > \rho$, even if an estimate of β' is less than unity. A low value of ρ implies that there is little correlation between initial and final values and hence a large mobility through the distribution. We must then be careful when interpreting the results of the β-convergence regressions and use the information given by the analysis of σ-convergence.

Which Convergence Notion?

Bernard and Durlauf (1996) show that cross-sectional tests used in this work—σ and β-methologies—may be interpreted with respect to the first definition of convergence. Nevertheless, they show that these tests fail to provide evidence for convergence compatible with the second definition. Indeed, the tests tend to spuriously reject the null hypothesis of no convergence even when the variables of interest exhibit different steady states. This can occur if economies converging to low-output equilibria start far below their steady states, whereas economies converging to high-output equilibria start at values near their steady states.

This observation calls into question the usefulness of cross-sectional tests in assessing theoretical disputes in the growth literature. However, the first definition seems to be the most useful from the perspective of this empirical work. The Maastricht convergence concept is less restrictive than the second definition. The only relevant nominal criterion to judge if a country can join the EMU is whether it has come closer to EMU members over an horizon of a few years. Hence the criterion is a middle-run criterion and does not consider the (infinite) long-run.

3. Real and Nominal σ-Convergence

The data are taken from various sources. The index of consumption prices comes from the IMF database. The real GNP figures are extracted from the Summers and Heston (1991) dataset. I consider all the members of the European Union and EFTA, except for Greece.[6] Taking Greece into account would greatly alter estimations for the nominal convergence without appreciably affecting the results for real convergence. The very high level of Greek inflation rates and their fluctuations introduce a bias in favor of the divergence hypothesis. Furthemore, all countries having the same weight in cross-sectional methods, Greece would have too important an effect on the results.

I start with a study of the path followed by the two processes using temporal graphical representations of univariate cross-sectional variances. I then move on to study the simultaneous evolution of nominal and real convergence.

Study of Univariate Cross-Sectional Variances

Consider the evolution of cross-country variances for per capita GNP. Figure 1 shows that there has been a global convergence over the entire period, but with an increased dispersion in the early 1980s. This seems to coincide with "restrictive" policies until the oil counter-shock. This temporary increase in the dispersion is mainly due to the behavior of the countries in the periphery of the Union.[7] For the first six members of the Community—Germany, France, Benelux, and Italy—the convergence has just slowed down during this period.

The dispersion of inflation rate cross-variances was very stable in the 1960s (Figure 2). Nevertheless, a constant variance can hide a crossed dynamic in the internal distribution, and the variance stability cannot be interpreted as reflecting inflation rate stability for the entire population. The oil shock in the 1970s brought with it a very important increase in the dispersion. In the second half of the 1980s we can note a convergence movement, despite the sudden increase of the variance in 1989–90. This increase can be essentially imputed to the behavior of Sweden, Germany, and Portugal. The 1990s are clearly marked by a large decrease of the dispersion.

A first look at these figures suggests some conclusions on the link between nominal and real convergence, but it is difficult to compare them, to see their similarity or their differences. One contribution of this paper is therefore to connect in a common figure, nominal convergence with real convergence.

Real Versus Nominal Convergence

The σ-convergence hypothesis can be represented by the condition $\sigma_{y,t+8}/\sigma_{y,t} < 1$, where $\sigma_{y,t}$ is the standard error of a variable y at date t.

The length of eight periods must be close to the length of a rolling filter, where we smooth the average increase rates. The convergence being a long-term process, the choice of this interval takes into account the length of the main economic cycles. On the other hand, the requirement to have a large number of points, in order to analyze the rolling evolution of the ratio, points to having a filter that is short enough. The length of eight years seems a good compromise.

For the present purposes we are justified in choosing the rolling estimate over the recursive estimate. The convergence being a long-term process, the temporal dimen-

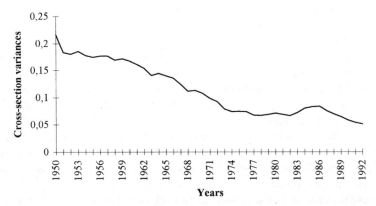

Figure 1. The Evolution of Cross-Country Variances for Per Capita GNP

Figure 2. The Evolution of Cross-Country Variances for Inflation Rate

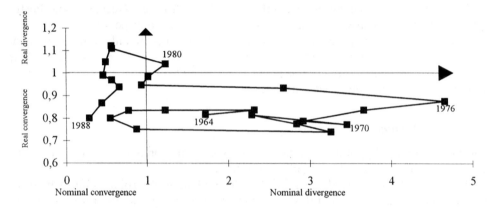

Figure 3. Convergence Path for Inflation Rates and Per Capita GNP

sion of samples plays an important role. Modifications occurring over time can be hidden by the regular increase of the estimation period length.[8] We need a constant period length to make comparisons easy. Moreover, the recursive method—which deals with the evolution of the ratio $\sigma_{y,t}/\sigma_{y,t_0}$—leads to results that strongly depend on the base year. As a matter of fact, if the dispersion of the variables is extremely narrow during the reference year (this is the case for this exercise), the recursive results would be biased for the whole period in favor of the divergence hypothesis. A rolling measure can deal with this problem.

I therefore chose to use the rolling method, despite the fact that it makes difficult the identification of possible breaking points. Indeed, as this method attaches the same weight to each observation, it is impossible to know if the change in the evolution of the ratio could be imputed to the removal of the rolling sample's initial data or to the introduction of the final data.

In order to make comparisons easier, we draw a figure cut out in four distinct zones, delimiting the areas in which nominal/real convergence occur. Thus, the four areas identified can be represented graphically, where the ratio σ_{t+8}/σ_t is affected to the date $t + 4$ (this amounts to centering the data on each interval).

We can see important modifications in the relationship between nominal and real convergence over time (Figure 3). At the beginning of the 1960s, there is a double

convergence movement. Conversely, we observe a period in the 1970s, of real convergence and nominal divergence. The beginning of the eighties (80–84) reveals real
divergence and nominal convergence. The end of the 1980s is characterized by the
return of a double convergence movement.

One might be tempted to conclude that the nominal convergence observed during
the 1980s occurred at the price of a per capita GNP divergence. This movement
coincided with the creation of the European Monetary System. The deflationary
policies seem to have ensured for this period nominal convergence to the detriment of
real convergence. However, it is difficult to draw any definite conclusions on the basis
of only these figures. We can hardly talk about links and correlations when the
observed movements could well be entirely coincidental. It is therefore interesting
to study the bivariate system—nominal and real. Indeed, a weak point of the analysis
so far is that we cannot distinguish, with a simple analysis of the σ-convergence,
the different dynamics of the processes, their respective stability, and especially their
cross-effects.

4. Estimation of a Two-Dimensional System

We have seen in the previous section that it seems difficult to consider, on empirical
grounds, nominal and real convergence as separate processes. The next step in the
analysis is therefore to study their joint dynamics. This is what we do in this section. I
first present the bivariate system and then report the results.

A Two-Dimensional System

Consider a two-dimensional system in which each convergence equation is conditioned
by the initial level of both nominal and real variables. This system can be written
(without noise) in the following way:

$$\begin{pmatrix} (x_T - x_0)/T \\ (y_T - y_0)/T \end{pmatrix} = \begin{pmatrix} a \\ b \end{pmatrix} + \begin{pmatrix} c & d \\ e & f \end{pmatrix} \begin{pmatrix} x_0 \\ y_0 \end{pmatrix},$$

where x is a real variable and y a nominal variable.

This joint dynamic can be analyzed as a representation of conditional convergence
with mutual conditioning. The model of conditional convergence predicts that countries can reach different stationary states: the convergence is conditioned by the determinants of the stationary state. Conditioning nominal (resp. real) convergence by the
initial level of the real (resp. nominal) variable, we can then study the potential
influences between the two types of convergence.

The case $e = d = 0$ corresponds to the unconditional convergence analysis. In
that case, nominal and real convergence processes are disjoint. This situation will
be called hereafter the "diagonal system," or "disjoint model," as opposed to the
"joint system," where e or d is different from 0. The diagonal system is equivalent
to the traditional β-convergence equation (see equation (1)). The coefficients c and f
are then estimated by the OLS method. Thus we can deduce from diagonal models the
estimated convergence speeds for the two types of variables. One gets an estimator
of λ:

$$\hat{\lambda} = -\ln\left(1 + \hat{h}T\right)/T,$$

where $\hat{h} = (\hat{c}, \hat{f})$ is the estimated coefficient of the univariate regression.

The particular interest of the joint-system characteristic equation is that it allows us to study different variable dynamics following the introduction of an initial gap between countries. The convergence will therefore be understood as a dynamic concept; i.e., the closing of initial gaps between countries. The basic system linking the average rate of increase, that is the average speed of a variable, with its initial level, can be rewritten so as to reveal the underlying dynamics. This simplified form without the noise can be expressed as in Cohen (1992):

$$dx_t/d_t = a + cx_0 + dy_0,$$
$$dy_t/d_t = b + ex_0 + fy_0.$$

Simple studying the signs of the coefficients in β-convergence models is not enough to conclude anything about the convergence of the variables. We must first analyze the stability of the system. Indeed, if the system were not stable, we would estimate conditional dynamics towards a moving target, and not convergence towards a stationary state. We will examine the stability of the system with the help of the usual criterion in linear dynamics: analysis of the trace and the determinant. As in Cohen (1992), we test if the determinants are positive with Wald's tests.[9]

Section 3 made clear that important changes occurred in the convergence processes. Thus, we will study these evolutions by comparing different periods. First we estimate a rolling evolution of the joint system. Second we study more deeply some periods chosen on historical grounds and estimate the coefficients for four time periods: from 1964 to 1973, from 1972 to 1979, from 1979 to 1987, and from 1987 to 1992. These sub-periods and the rolling periods are never exactly the same. Finding essentially the same results with rather different estimation periods increases the robustness of the conclusions.

Rolling estimations of the joint system   We cross-estimate, for each rolling sub-period, the two-dimensional system, where convergence movements are reciprocally conditioned. The coefficients of these regressions are reported in Figures 4, 5, and 6.

The coefficients, called "direct," are the estimated coefficient of inflation at its initial level in the nominal convergence equation, and the estimated coefficient of the initial per capita GNP level in the real convergence equation. Figure 4 shows the convergence or divergence tendencies of the two types of process. A more careful study is, however, necessary to reach a conclusion, because we need to check the stability of the system and the influence of cross-effects. We nevertheless observe in this first analysis the existence of four distinct periods:

from 1965 to 1971: nominal convergence–real convergence
from 1972 to 1978: nominal divergence–real convergence
from 1979 to 1984: nominal convergence–real divergence
from 1984 to 1988: nominal convergence–real convergence.

Recall at this point that these dates are centered on rolling samples. Thus, the estimation set, for example, on 1965 is the estimation achieved for the interval 1961–69. The confidence intervals are not shown on this figure: the coefficients are always significant, except around the two oil shocks for the nominal variable, and when divergence occurred for the real variable.

The results concerning the cross-coefficients, reported in Figures 5 and 6, show that the influence of the initial level of per capita GNP on the growth rate of inflation is more significant than the impact of the initial inflation level on the rate of the logarithm

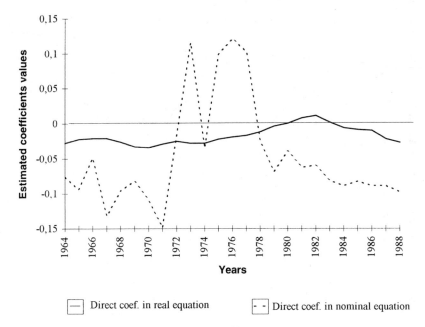

Figure 4. Rolling Estimates of the Direct Coefficients

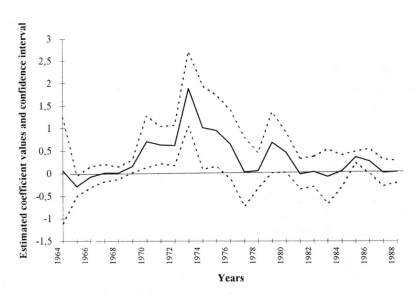

Figure 5. The Real Coefficient in the Nominal Convergence Equation

of per capita GNP. We hence note an asymmetry in the system. It is not the aim of this paper to test any particular theory, but we could maybe understand these results as indicating the presence of a significant Phillips effect in the price equation, and the absence of a competitive effect in the growth equation.

The coefficient of the initial per capita GNP level in the nominal convergence equation is in fact significantly positive on a set of adjacent intervals, which define the

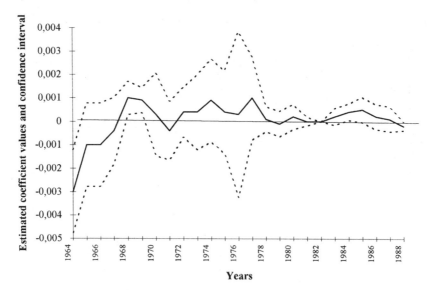

Figure 6. The Nominal Coefficient in the Real Convergence Equation

period 1973–80; i.e., from 1969 to 1976 on Figure 5. This period corresponds to the beginning of the divergence movement for the nominal variable up to the beginning of the divergence movement for the real variable. In this period, the weaker the initial per capita GNP of a country, the higher is the increased rate of per capita GNP, and the weaker is the inflation growth rate.

In fact, this effect ceased to be significant when real divergence occurred (at the beginning of the 1980s). In short, this evidence confirms the importance for this period of wage and price controls as a key component of the disinflation process. It suggests that the policy was not simply to rely on unemployment to restrain costs, as noted by Blanchard and Muet (1993).

Estimations for Periods Chosen on Historical Grounds

We next consider sub-periods, chosen on historical grounds: the break points are taken around the two oil shocks, and the oil counter-shock of 1986. An "independence" test[10] is made on each sub-period in order to identify situations in which we cannot reject the diagonal model in favor of the joint system. These situations reveal that nominal and real processes are independent.

Estimated coefficients of the diagonal model and of the joint system, for the four time periods, are presented together in Table 1. The χ^2 statistics associated with the independence test[11] lead us to reject the diagonal model for two out of these four periods. We thus focus on the study of the joint-system coefficients for all periods except 1979–87 and 1987–92. For these periods, we accept the null hypothesis and do not reject the diagonal model. It is then useless to study the joint system and the nonsignificant cross-effects. The convergence dynamics become dichotomic in course of time.

The sign of the coefficients linking the average increase rates to their own initial level is negative for the first period (1964–73). This reveals a movement of nominal and real convergence. Inflation rates converge about two times faster than real per capita

Table 1. Estimations of the Diagonal and Joint System

Period	System		Δp_0	$\ln y_0$	$\bar{R}^2$	λ
1964–73	Diagonal	$\Delta^2 p$	−0.052 (−2.57)		0.27	0.07
		$\Delta \ln y$		−0.03 (−8.19)	0.81	0.035
	Joint	$\Delta^2 p$	−0.051 (−3.12)	0.205 (2.25)	0.41	
		$\Delta \ln y$	0.001 (2.03)	−0.03 (−9.79)	0.84	
1972–79	Diagonal	$\Delta^2 p$	0.29 (1.9)		0.15	−0.158
		$\Delta \ln y$		−0.023 (−2.99)	0.35	0.025
	Joint	$\Delta^2 p$	0.287 (2.27)	1.035 (2.17)	0.29	
		$\Delta \ln y$	−0.001 (−0.47)	−0.023 (−3.21)	0.3	
1979–87	Diagonal	$\Delta^2 p$	−0.083 (−6.81)		0.75	0.136
		$\Delta \ln y$		0.003 (0.55)	−0.05	−0.0035
	Joint	$\Delta^2 p$	−0.082 (−6.63)	−0.063 (−0.23)	0.73	
		$\Delta \ln y$	0.0002 (0.83)	0.001 (0.23)	−0.08	
1987–92	Diagonal	$\Delta^2 p$	−0.15 (−4.99)		0.61	0.277
		$\Delta \ln y$		−0.041 (−3.42)	0.41	0.045
	Joint	$\Delta^2 p$	−0.159 (−5.66)	0.33 (1.13)	0.61	
		$\Delta \ln y$	−0.0004 (−0.36)	−0.04 (−3.47)	0.37	

Student's statistics for the coefficients are reported in parentheses. The λ parameter is the convergence speed, calculated in the univariate system.

standards of living. The system is stable[12] and all coefficients of the system are significant. The cross-effects reveal a negative contribution to the convergence, whose dynamics can be studied with the help of Figure 7. The impact of an initial gap in standard of living on nominal convergence is more important than the influence of the initial gap in inflation rate on real convergence.

Conversely, we observe (Figure 8) that in the 1972–79 period, nominal convergence has ceased. The coefficients are both significantly positive in both univariate and

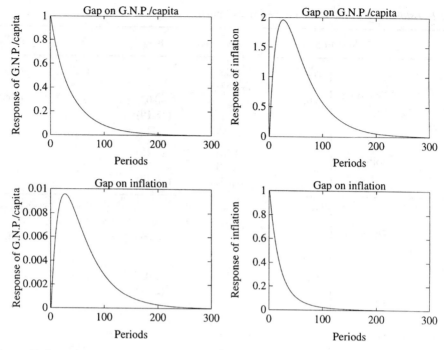

Figure 7. Impulse Response Functions Following the Introduction of an Initial Gap Between Countries (1964–73)

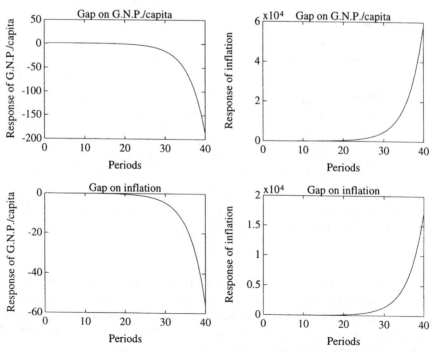

Figure 8. Impulse Response Functions Following the Introduction of an Initial Gap Between Countries (1972–79)

bivariate analyses. The real convergence, however, continues, although at a lower speed. We can note the importance of the initial per capita GNP level in nominal convergence. If a country has a high initial standard of living, it will experience higher average increased rates of inflation. Statistical analysis of the determinant and trace[13] reveals that the equilibrium is unstable, with the possibility of a saddle-point. The diverging influence of the nominal variable and the explosive character of the system explain the real diverging movement in the long run.

For the 1979–87 period, the situation is reversed. Once again, nominal convergence occurs at the cost of real convergence. The only significant coefficient links the initial inflation rate level to its own average increased rate. We then observe a nominal unconditional convergence movement, that could be called dichotomic. The equilibrium appears unstable with the possibility of a saddle-point.[14] This sub-period is the first for which the real initial level has no more influence on nominal convergence, and coincides with the period in which real convergence is no longer realized. The sustained deflationary effort and a slowdown in the real convergence process arose at the same time.

Lastly, we see again, for the 1987–92 period, a double convergence movement: both direct coefficients are significantly negative. The convergence speed is very high. Study of the trace and the determinant reveal the stability of the joint system.[15] As opposed to 1964–73, the convergence processes become for this last period relatively dichotomic. An initial gap in per capita GNP involves a weaker nominal reaction. Moreover, the closing of the gap is performed faster.

4. Conclusion

The aim has been to study nominal and real convergence, as reciprocally conditioned within the same system. The joint dynamics of inflation and GNP growth were considered as a framework to test for an interrelation between nominal and real convergence among European countries.

Preliminary evidence based on σ-convergence analysis revealed that nominal and real convergence were negatively correlated in the 1980s. Contrary to what happened during the 1970s, nominal convergence occurred in the 1980s at the cost of real divergence. This negative correlation disappeared in the 1990s. Thus, at the end of the sample period both a nominal and a real convergence movement were observed.

Estimation of a bivariate system exposes the mutual relationships between these different movements. The initial per capita GNP level influenced the nominal convergence in the 1970s. However, this effect ceased to be significant when real convergence slowed down at the beginning of the 1980s. In the 1990s we see a simultaneous convergence movement, the two cross-effects no longer being significant. It seems, then, difficult to say that the nominal convergence leads to real divergence. The convergence dynamics have become dichotomic, and nominal convergence seems no longer to interfere with real convergence.

References

Ardeni, Pier G., "On the Way to the EMU Testing Convergence of the European Economies," *Economic Notes by Monte del Paschi di Sierra* 21 (1992):238–57.
Barro, Robert, "Economic Growth in a Cross-Section of Countries," *Quarterly Journal of Economics* 106 (1991):407–43.
———, "Inflation and Economic Growth," NBER working paper 5326, 1995.

Bernard, Andrew B. and Steven N. Durlauf, "Interpreting Tests of Convergence Hypothesis," *Journal of Econometrics* 71 (1996):161–73.

Blanchard, Olivier J. and Pierre A. Muet, "Competitiveness Through Disinflation: An Assessment of the French Macroeconomic Strategy," *Economic Policy* (1993):12–44.

Cohen, Daniel, "Tests of the Convergence Hypothesis: a Critical Note," manuscript, Cepremap, 1992.

Feldstein, Martin, "Europe's Monetary Union: the Case against EMU," *The Economist* (1992):19–22.

Gouriéroux, Christian and Alain Monfort, *Séries Temporelles et Modèles Dynamiques*, Paris: Economica, 1990.

Haldane, A. G. and S. G. Hall, "Sterling's Relationship with the Dollar and the Deutschmark: 1986–89," *Economic Journal* 101 (1991):436–43.

Hall, S. G., D. Robertson, and M. R. Wickens, "Measuring Convergence of the EC Economics," *The Manchester School* LX Supplement (1992):99–111.

Hart, Peter, "Galtonian Regression Across Countries and the Convergence of Productivity," *Oxford Bulletin of Economics and Statistics* 57 (1995):287–93.

Hénin, Pierre Y. and Yannick Le Pen, "Les épisodes de la Convergence Européenne," *Revue Economique* 46 (1995):667–77.

Heylen, Freddy, André Van Poeck, and Johan Van Gompel, "Real Versus Nominal Convergence: National Labour Markets and the European Integration Process," *Labour* 9 (1995):97–119.

Lichtenberg, Franck R., "Testing the Convergence Hypothesis," *Review of Economics and Statistics* (1994):576–9.

Quah, Danny, "Galton's Fallacy and Tests of the Convergence Hypothesis," *Scandinavian Journal of Economics* 95 (1993):427–43.

———, "Empirics for Economic Growth and Convergence," *European Economic Review* 40 (1996):1353–75.

Summers, Robert and Alan Heston, "The Penn World Table: An Expanded Set of International Comparisons 1950–1990," *Quarterly Journal of Economics* (1991):327–68.

Notes

1. However, statistically significant results emerge only when high-inflation experiences are included in the sample.

2. This is the nominal variable specified by the Maastricht Treaty. However, in the long run, in a fixed exchange rate system such as the EMU, convergence of inflation rates is not enough, and that of price levels is required.

3. σ_t is the (N-sample-size) cross-sectional standard deviation:

$$\sigma_t = \left(N^{-1} \sum_{i=1}^{N} \left[y_{i,t} - \left(N^{-1} \sum_{k=1}^{N} y_{k,t} \right) \right]^2 \right)^{1/2}.$$

4. Hénin and Le Pen (1994) illustrate this discussion in an exercise, where they demonstrate how a simple diagrammatic representation may show the distinction between σ- and β-convergence.

5. Indeed $\beta' = \text{cov}(y_{i,t+1}, y_{i,t})/\sigma_t^2$ and $\rho = \text{cov}(y_{i,t+1}, y_{i,t})/\sigma_{t+1}\sigma_t$.

6. More precisely, we consider the following countries: Austria, Belgium, Denmark, Eire, England, France, Finland, Germany, Italy, Luxembourg, The Netherlands, Norway, Portugal, Spain, Sweden, Switzerland.

7. We note especially the behavior of Spain and Portugal.

8. This problem would be even more serious in the next section when we estimate a β-convergence system. With a recursive analysis, the coefficient depends on the parameter T, which measures the length of the estimation period. Therefore, the convergence speed and the $\bar{R}^2$ of the regressions are very sensitive to the changes of the period T.

9. In order to know whether we can reject the hypothesis H_o that $\Theta = bc' - b'c > 0$, we perform the following test, presented by Cohen (1992). Let $u = (\partial\Theta/\partial b, \partial\Theta/\partial c, \partial\Theta/\partial b', \partial\Theta/\partial c')$, so $u = (c', -b', -c, b)$. Call $\hat{\Sigma}$ the covariance matrix of the estimated coefficients $(\hat{b}, \hat{c}, \hat{b}', \hat{c}')$, and note $\hat{V}(\hat{\Theta}) = \hat{u}\hat{\Sigma}\hat{u}^T$. We calculate the student statistics of Θ: $\varepsilon = \hat{\Theta}/\hat{V}(\hat{\Theta})^{1/2}$.

10. We recall here the test presented by Gouriéroux and Monfort (1990, p. 604). Let $Z_t = AW_t + u_t$ for $t = 1, \ldots, T$ a linear multivariate regression model, where Z_t is an s-vector, W_t a k-vector, A a $(s \times k)$ matrix of unknown parameters, and u_t a white noise. Calling H_o the hypothesis on A, the likelihood ratio statistic is: $\varepsilon_T^R = T \log (\det\hat{\Omega}_o/\det\hat{\Omega}) \to \chi^2_{1-\alpha}(r)$, where $\hat{\Omega}_o$ and $\hat{\Omega}$ are the maximum likelihood constrained and unconstrained estimators of Ω, r the constraint's number, and α the hold level.

11. The χ^2 statistic associated with the independence test is about 8.23 for the 1964–73 time period, 4.47 for the 1972–79 period, 0.737 for the 1979–87 period, and 1.36 for the last period.

12. The determinant is, for this period, equal to 0.00132, and Wald's test allows us to accept the assumption that the determinant is positive at the 5% level. The trace is equal to –0.081 and is significantly negative at the 1% level. The student's statistic associated with the trace is –4.87.

13. The determinant is then negative (–0.00541) and Wald's test allows us to reject the assumption that the determinant is positive. The trace is equal to 0.257 and its Student statistic is 2.26. It is significantly positive at the 5% level.

14. The trace (–0.081) and the determinant ($-6.94.10^{-5}$) are both negative.

15. The trace is significantly negative at the 1% level and the determinant is positive. Wald's test allows us to accept the assumption that the determinant is positive. Student's statistic associated with the trace is –6.62.

Review of International Economics, Special Supplement, 92–110, 1997

Employment and Wage Bargaining in an Open Monetary Union

Pierre Cahuc and Hubert Kempf*

Abstract

This paper studies alternative patterns of wage bargaining in an open two-country monetary union. Wages are fixed by trade unions for two periods, either at the national or at the monetary union level. It is shown that the best solution with regard to unemployment depends on the nature of externalities and dynamic strategic interactions between the monetary union's countries; namely on the degree of openness of the monetary union, and the differentiation index between national goods.

1. Introduction

The prospect of a common currency for several European countries radically alters the issue of economic relationships and internal adjustments among the countries belonging to such a monetary union: there would be no need to balance trade flows between these countries by means of exchange rates, and transaction costs linked to currency changes would be suppressed. A forceful argument in favor of a European common currency indeed stresses the need to make the relationships between the member countries immune to exchange rate variations and risks, in order to neutralize a major source of uncertainty, if a closer economic integration of these countries is wanted.

The issue of internal adjustments within a monetary union would be trivial if all real and financial markets, both within the monetary union and outside it, could be considered as perfectly competitive markets. The reality of imperfect competition makes it much more difficult to address. In this paper, we study the functioning of a two-country monetary union[1] (hereafter MU), open to the external world, when some features of imperfect competition are taken into account, to show how these characteristics affect the adjustment patterns within the MU.[2]

Wage bargaining is an important aspect of imperfect competition and has been studied extensively. Driffill and van der Ploeg (1993) studied the impact of a unique market on wage bargaining, but did not consider the possibility of a common currency, nor the effects of external competition on wage bargaining. The present paper represents a first step towards an understanding of the consequences of wage bargaining in an open monetary union. It is assumed that wages are not fixed in a perfectly competitive labor market, but negotiated by trade unions within a monetary union of interdependent economies, sharing the same currency, and open to the world economy. Apart from the existence of trade unions, we will also assume the existence of another source of imperfect competition, that is wage contracts covering several periods. This charac-

---

* Cahuc: MAD, Université Paris-1 90 rue de Tolbiac, F-75013 Paris, France. Tel: (33)-(01)-40771912; Fax: (33)-(01)-45847889; Email: cahuc@univ-paris1.fr. Kempf: MAD, Université Paris-1 Panthéon-Sorbonne, 90 rue de Tolbiac, F-75013 Paris, France. Email: kempf@univ-paris1.fr. This paper was partially written while the second author was Jean Monnet fellow at the European University Institute, Florence. We are grateful to Pierre-Yves Hénin for useful discussions at the beginning of this research project. We would like to thank participants to seminars at MAD, Université Paris-1, EUI (Florence), IMOP (Athens), and especially Huw Dixon, Tryphon Kollintzas and an anonymous referee, for very helpful comments made on a previous version of the paper. Any remaining errors are ours.

teristic is found, in various degrees, on every labor market in developed countries. On theoretical grounds, the notion of multiperiod wage contracts is well-established, particularly in international macroeconomics (Fischer, 1977; Canzoneri and Henderson, 1991).

Within such a framework, it is worth studying the consequences of different wage settings according to two criteria: the optimal level of bargaining, and the time pattern of negotiation. The last point has been studied less than the first. Calmfors (1993) surveyed the results that have been obtained on the first issue (see also Calmfors and Driffill, 1988; Cahuc and Zylberberg, 1991). De Fraja (1993) compared the outcome of staggered and synchronous wage setting in a closed economy when wages are set at the firm level. The originality of our paper is to provide a simple framework that allows a comparison of the outcomes of different wage settings in a monetary union, mixing these two dimensions.

Here we analyze the importance of alternative time patterns of wage bargaining using the concept of Markov Perfect Equilibrium (MPE) developed by Maskin and Tirole (Maskin and Tirole, 1988a, 1988b, 1989). In another paper (Cahuc and Kempf, 1995) we have shown that the impact of different time patterns of decisions depends upon both the nature of externalities and the type of strategic interactions linking agents. Applying this reasoning to the present framework, we will show that it is possible to compare the stationary solutions corresponding to an MPE in the different dynamic wage settings we consider. More precisely, the ranking between the wage and employment levels depends, through a simple relationship, on the openness index of the union and the level of differentiation between the goods produced in the union. From this we can infer whether or not wage bargaining should be centralized at the union level. Moreover, in the case of noncooperative behavior between trade unions, we are able to compare the consequences of decentralization, when the timing of wage negotiation is taken into account, and to endogenize the time pattern of trade unions' decisions. Lastly, we show that our results hold whether monetary policy is discretionary or follows a constant money supply rule.

Clearly, the model presented in this paper is specific in some ways. In particular, we do not allow for uncertainty which would have led us to consider indexation issues as in Fethke and Policano (1984),[3] Spivak et al. (1987), and Kempf (1989). However, since our ranking between bargaining patterns is strict, it follows that a small amount of uncertainty is unlikely to reverse the ranking obtained. On the whole, we hope that this paper will shed some light on institutional arrangements relative to wage-setting within a monetary union.

In section 2 we set up a model of a two-country monetary union, open to the rest of the world, where wages are set for two periods by trade unions, allowing us to consider three possible variants differing according to the time pattern and the level of the bargaining structure. These three cases are then considered in section 3, so that a comparison of the steady-state wage and employment levels can be achieved.

2. The Model

We consider a world economy consisting in three countries. On the one hand, there are two countries with the same structure and size, sharing a common currency (the euro), and hence forming a monetary union. There is one country representing the rest of the world. There are flexible exchange rates linking the MU and the rest of the world. The MU is of negligible weight relative to the rest of the world. The MU's current account balance is always cleared.

Each country belonging to the MU produces a differentiated good. These two goods, good A produced by country a and good B produced by country b, are imperfectly substitutable and can be bought by consumers in the rest of the world as well as by consumers in the MU. Moreover, the rest of the world is producing a third good, good X, purchased by consumers in the MU. P_a, P_b, and P_x represent the prices (in euros) of goods A, B, and X. E denotes the exchange rate between the MU and the rest of the world, and P_x^* is the price, exogenously determined of good X, expressed in foreign currency. Hence, we may write

$$P_x = EP_x^*. \tag{1}$$

Money supply by the central bank is denoted M^0; we will suppose it is constant and equal to $\bar{M}$. At the end of the paper, we will relax this assumption and study the consequences of endogenizing money supply. Since the current account is balanced (and money balances remain constant), we may write

$$P_a Y_a + P_b Y_b - P_a C_a - P_b C_b = EP_x^* C_x = P_x C_x, \tag{2}$$

where Y_a and Y_b are the quantities of goods A and B produced within the MU, and C_a, C_b, and C_x are the quantities of goods A, B, and X purchased by MU's consumers.

The Supply of Goods

In each country belonging to the MU, there is a large number of perfectly competitive firms. The sole production factor is labor. Since returns are constant, the price of good i is equal to the wage rate currently existing in country i:

$$p_i = w_i, \quad i = a, b \tag{3}$$

(capital letters correspond to levels and lower-case letters to the logarithms of corresponding variables). Similarly, it is assumed that the price of good X equals marginal cost:

$$p_x^* = w^*, \tag{4}$$

where w^* is the (log of the) nominal wage rate in the rest of the world.

The MU being negligible compared with the rest of the world, the wage rate W^* currently in use in the rest of the world is independent of the wage behavior in the MU. Consequently, the price level of good X, P_x^*, may be considered as fixed.

Aggregate Demand for the Various Goods

All consumers from the MU are identical. The representative consumer is characterized by the following utility function:

$$U(C_u, C_x, M/P) = C_u^{\alpha(1-\lambda)} C_x^{\alpha\lambda} (M/P)^{1-\alpha}, \quad 0 < \alpha < 1, \quad 0 < \lambda < 1, \tag{5}$$

where C_u is the representative agent's consumer basket defined on goods A and B, and C_x its consumption of the imported foreign good. M/P represents the consumer's real balances, P being the MU's consumption price level, defined as

$$P = P_u^{1-\lambda} P_x^\lambda. \tag{6}$$

The coefficient λ represents the weight of foreign good consumption relative to the consumption of MU's goods within the agent's preference. It may be interpreted as an openness index of the MU as a whole on the rest of the world.

The representative agent's "internal" consumption basket is defined as

$$C_u = (2)^{\frac{1}{1-\theta}} \left[C_a^{\frac{\theta-1}{\theta}} + C_b^{\frac{\theta-1}{\theta}} \right]^{\frac{\theta}{\theta-1}}, \quad \theta > 0, \ \theta \neq 1, \tag{7}$$

where θ represents the elasticity of substitution between A and B. θ formalizes the degree of differentiation between the two countries belonging to the MU. When θ is less than 1, goods A and B are gross complements, and when θ is greater than 1 they are gross substitutes. The bigger θ, the more substitutable they are. We will therefore define θ as a differentiation index between goods. Consequently, the MU's production price index P_u is given by the following formula:

$$P_u = \left[\left(P_a^{1-\theta} + P_b^{1-\theta} \right) / 2 \right]^{\frac{1}{1-\theta}}, \tag{8}$$

and we can write

$$P_u C_u = P_a C_a + P_b C_b. \tag{9}$$

All outside consumers are identical and have the same utility function as the MU's consumers. In particular, they have the same preference structure towards the goods produced by the MU's countries as the MU's consumers.

For the MU, taken as a whole, the (average) aggregate income is

$$R \equiv W_a L_a + \Pi_a + W_b L_b + \Pi_b = P_a Y_a + P_b Y_b, \tag{10}$$

where W_a, W_b, L_a, L_b, Π_a, Π_b are respectively the wage rates, the quantities of labor, and the profits of a representative firm of a and b.[4] The representative agent's budget constraint is written

$$P_u C_u + P_x C_x + M = \overline{M} + R, \tag{11}$$

and the demand functions for the (basket of) internal and external consumption deduced from (5) and (11) are

$$P_u C_u = \alpha(1-\lambda)(\overline{M} + R),$$
$$P_x C_x = \alpha\lambda(\overline{M} + R),$$
$$M = (1-\alpha)(\overline{M} + R).$$

Equilibrium on the money market implies

$$M = \overline{M}. \tag{12}$$

The consumer's demand function implies

$$P_u C_u = \left[\alpha(1-\lambda)/(1-\alpha) \right]\overline{M}, \quad P_x C_x = \left[\alpha\lambda/(1-\alpha) \right]\overline{M}. \tag{13}$$

The demand for good U addressed to the MU is given.[5] We can therefore write

$$P_u^* C_u^* = D^*, \tag{14}$$

where C_u^* denotes the import by the rest of the world from the MU of the index good U, and P_u^* denotes its price expressed in foreign currency. The global import of good

U by the rest of the world corresponds to a constant share of their total expenditures thanks to the consumer preferences we use; and through a reasoning similar to the one we used for the MU, $D*$ is proportional to the money supply in the rest of the world, assumed to be exogenous. Since the balance of payments is cleared, we can write

$$E = P_x C_x / D*. \tag{15}$$

Hence, from (13) and (14), according to preceding equalities:

$$E = \left[\alpha\lambda/(1-\alpha)\right]\overline{M}/D*. \tag{16}$$

The exchange rate is therefore a linear function of the ratio between money supplies in the MU and in the rest of the world. It depends neither on the individual prices of the goods nor on the wage rates in the MU and in the rest of the world, and may be considered exogenous. Also, given (1), P_x does not depend on the MU's wage rate. Hence, we will consider it as exogenous throughout the paper.

Furthermore, total nominal production $P_u Y_u$ (equal to total income, since external flows are cleared) is equal to the sum of internal and external consumptions of the MU. From (13), the demand functions of goods *A* and *B* are (Blanchard and Fischer, 1989, pp. 375–9):

$$Y_i = \left[\alpha(1-\lambda)/(1-\alpha)\right]\left(\overline{M}/P_u\right)\left(P_i/P_u\right)^{-\theta}, \quad i = a, b. \tag{17}$$

We will approximate the internal price level for the MU in the neighborhood of the symmetric equilibrium by means of the following relation in logs:[6]

$$p_u = \left(p_a + p_b\right)/2. \tag{18}$$

The demand function for good *i* can then be written

$$y_i = c + \overline{m} + \left(\theta - 1\right)p_u - \theta p_i, \quad i = a, b, \tag{19}$$

where *c* is a constant. We get, using the definition of p_u and omitting *c*:

$$y_i = \overline{m} + \left[\left(\theta - 1\right)p_j - \left(\theta + 1\right)p_i\right]/2, \quad i = a, b; \ j = a, b; \ j \neq 1. \tag{20}$$

3. Alternative Time Patterns of Wage Bargaining and Markov Perfect Equilibria

The instantaneous payoff function for the trade union in a given country depends on two arguments: *negatively* on the (squared) gap between actual and a target (natural) level of employment, i.e., of production; and *positively* on the real wage rate. In other words, a trade union desires to achieve two goals: a target employment level, together with the highest possible real wage level. The payoff function is the same for both trade unions, given by the following formula:

$$v\left(y_i, w_i, p\right) = -\left[\left(y_i - \overline{y}\right)^2/2\right] + \gamma\left(w_i - p\right), \tag{21}$$

where $\gamma > 0$ is the weight of the wage rate objective relative to the employment objective in the payoff function.

The MU's workers have the same preference as its consumers and the deflator of the nominal wage is the general price level in the MU. According to equations (5) and (11), it is defined as follows:

$$p = (1-\lambda)p_u + \lambda p_x = \left[(1-\lambda)(p_a + p_b)/2\right] + \lambda p_x. \tag{22}$$

The interdependence between trade unions comes from the fact that workers from a given country consume the good produced in the other country, as well as the good X produced in the rest of the world. In a purely static framework (abstracting from any dynamic consideration), if trade union b increases the wage rate prevailing in country b, this increases the price of good B, depressing the real wage rate in country a. But at the same time, the rise in nominal wages in country b increases the desirability of good A at a given wage rate both within the MU and in the rest of the world, and this is beneficial for employment. These two effects being opposite, we cannot know *a priori* the sign of the net externality between countries.

We can rewrite (21) from (20) and (3):

$$v(w_i, w_j) = -(1/2)\left\{\overline{m} - \overline{y} - \left[(\theta+1)w_i + (\theta-1)w_j\right]/2\right\}^2$$
$$+ \gamma\left[(\theta+1)(w_i/2) + (\theta-1)(w_j/2) - \lambda p_x\right]. \tag{23}$$

Note that the instantaneous payoff function, being quadratic, is concave since $v_{11} + 2v_{12} + v_{22} < 0$.

Both trade unions have the same discount factor $\delta \in \,]0,1[$ and their temporal horizon is infinite. The wage bargaining procedures in the economy, given the assumptions we have just made, depend on two factors.

The bargaining level   Bargaining may take place separately in each country in the MU. We will say in this case that it is *decentralized* and that trade unions negotiate noncooperatively. We will then assume that each trade union adopts a Markov strategy. That is, when it has to take a decision (to choose the wage rate to be applied in its country), it only takes into account the other trade union's currently prevailing wage. In other words, the trade union only reacts to the decision of the other's country/trade union which directly affects its instantaneous payoff function (Maskin and Tirole, 1988a). We will then look for the Markov perfect equilibria of the game (i.e., the pairs of subgame-perfect Markov strategies). Another possibility is that bargaining takes place at the level of the MU, and we will then say that bargaining is *centralized*. We may consider that there exists a coalition of trade unions setting the wage rate for every firm in the MU. The solution will then correspond to the cooperative (within the MU) solution of the game.

The bargaining time pattern   The issue of the time pattern of bargaining derives from the two-period wage contracts. We consider two cases. The first case is when both trade unions are synchronized, setting nominal wages simultaneously (let us say at the beginning of each even period). The second case is when they may alternate or stagger their decision, so that one trade union takes its decision at the beginning of each even period and the other one at the beginning of each odd period.

The Case of Synchronized and Decentralized Bargaining

Trade union *i*'s intertemporal payoff function may be written as

$$V_{i,t} = \sum_{t=0}^{\infty} \delta^t v(w_{it}, w_{jt}), \quad i = a, b; \ j = a, b; \ i \neq j. \tag{24}$$

Wages being fixed for two periods, trade union *i*'s problem may be written as

$$\max_{w_{it}} V_{i,t} = v(w_{it}, w_{jt}) + \delta v(w_{it}, w_{jt}) + \delta^2 V_{i,t+2}.$$

Since we are looking for stationary solutions for the game, and since $V_{i,t+2}$ does not depend on W_t, this problem is equivalent to

$$\max_{w_i} (1 + \delta) v(w_i, w_j). \tag{25}$$

The first-order condition implies $v_1(w_i, w_j) = 0$. Hence:

$$w_i = \left[(\theta - 1)/(\theta + 1)\right] w_j + \left[2/(\theta + 2)\right]\left[\overline{m} - \overline{y} + \gamma(1 + \lambda)/(\theta + 1)\right]. \tag{26}$$

We can easily deduce from (26) the reaction function of trade union *i* to the action taken by trade union *j*. The reaction function of a trade union in the synchronized case is denoted $R^S(\cdot)$. From the first-order condition, we get

$$R^{S'}(w_j) = 1 + (\theta - 1)/\theta. \tag{27}$$

Following Cooper and John (1988), there is strategic *complementarity* between players when the reaction function $R^S(\cdot)$ is increasing, and strategic *substitutability* when it is decreasing. The sign of dynamic strategic interactions between trade unions depends on the value of θ. From (26), the strategic effect between players comes from the output argument in their instantaneous payoff function. Player *j*'s wage intervenes in the marginal variation of player *i*'s welfare as an argument in the output term of the welfare function. When both goods *A* and *B* are substitutable, when a given trade union increases the nominal wage of its workers, the other one is induced to increase its own nominal wage, since the marginal welfare gain of a wage increase is positive. The explanation is as follows. For a given wage pair, an increase in *j*'s nominal wage induces trade union *i* to increase its own wage rate: *j*'s wage rise increases output and hence worsens *i*'s welfare. But, because of the goods substitutability in the consumer market, an increase in *i*'s wage rate helps to decrease output relative to its target (natural) level and (partially) restore *i*'s welfare. The reverse effect happens in the case of goods complementarity.

The stationary solution corresponding to decentralized and synchronized bargaining, which we denote w_S, is equal to

$$w_S = \overline{m} - \overline{y} + \gamma(1 + \lambda)/(1 + \theta). \tag{28}$$

The second-order condition $v_{11}(w_i, w_j) < 0$ is satisfied since, from the definition of $v(w_i, w_j)$ (cf. (23)):

$$v_{11} = -(1 + \theta)^2/4. \tag{29}$$

Note that for the decentralized and synchronized solution, we obtain the following value for the externalities between both players:

$$v_2(w_i, w_j) = -\left[(\theta - 1)/2\right]\left[\overline{m} - \overline{y} - (\theta + 1)(w_i/2) + (\theta - 1)(w_j/2)\right] - \gamma(1 - \lambda)/2. \tag{30}$$

which implies

$$v_2(w_S, w_S) > 0 \Leftrightarrow (\theta-1)(1+\lambda) > (\theta+1)(1-\lambda) \Longleftrightarrow \lambda > \theta^{-1}. \tag{31}$$

For the noncooperative synchronized value of the wage rate, there is positive externality between both national trade unions when the openness index is bigger than the inverse of the differentiation index (i.e., the market power index of trade unions). If the MU is open enough to the rest of the world, the negative effect of an increase in country j's nominal wage has a small price effect within the MU, and hence has a small effect on country i's real wage rate. However, if both goods produced in the MU are weakly differentiated, this wage increase allows country i to win market share, and hence increases country i's employment. On the other hand, if the MU is closed enough and the price effect large enough, then the net welfare effect can be negative.

The Case of Staggered and Decentralized Bargaining

Consider now the case where trade union i bargains at the beginning of even periods and trade union j at the beginning of odd periods. The discounted intertemporal value of trade union i's payoffs when it bargains is

$$V_i(w_j) = \{v(w_i, w_j) + \delta W_i(w_i)\}, \quad i = a,b; \; j = a,b; \; i \neq j, \tag{32}$$

with

$$W_i(w_i) = v(w_i, R^{jl}(w_i)) + \delta V_i(R^{jl}(w_i)),$$

$$R^{il}(w_j) \in \arg\max_{w_i}\{v(w_i, w_j) + \delta W_i(w_j)\},$$

where $V_i(w_j)$ represents the discounted intertemporal value of trade union i's payoffs, computed for an even period, when trade union i takes its decision reacting to the decision w_j of trade union j, taken in the previous period; $W_i(w_i)$ represents the discounted intertemporal value of trade union i's payoffs, computed for an odd period, i.e., when trade union j takes its decision, reacting to the decision w_i of trade union i; and $R^{il}(w_j)$ represents the reaction function of trade union i to an action w_j taken by trade union j, in the case of alternate bargaining.

We assume that the reaction function of a given trade union is linear:

$$w_i = R^{il}(w_j) = \alpha_i + \beta_i w_j. \tag{33}$$

We then need the following lemma to establish the main proposition. It is proved in the Appendix.

LEMMA 1. *There exists a unique wage rate value (which we denote* w_1*) corresponding to a stable Markov perfect equilibrium of the staggered game for δ sufficiently high.*

A sufficient condition for a unique solution corresponding to the staggered pattern of bargaining is simple: the discount factor should not be too small; or equivalently, both trade unions should not have too strong a preference for the present, so that they are willing to take into consideration the future consequences of their current action on their future welfare.

The Case of Synchronized and Centralized Bargaining

Consider now the case where the prevailing wage rates in both countries belonging to the MU are fixed (always for two periods) by a coalition of trade unions covering the whole MU. As both countries are identical, it fixes the same wage rate for all firms and workers. The wage-setting being centralized at the MU level, the MU's coalition is able to internalize any direct or indirect effect linked to the wage rate and the competition on the goods markets of the MU's firms: it is able to set the cooperative solution between both countries.

Given the instantaneous payoff function, the centralized trade union's maximization problem is written as

$$\max_{w}(1+\delta)\upsilon(w,\ w) = (1+\delta)\left\{-\left[(\overline{m}-w-\overline{y})^2/2\right]+\gamma\lambda(w-p_x)\right\}. \tag{34}$$

From the first-order condition, we get

$$w_C = \overline{m} - \overline{y} + \gamma\lambda, \tag{35}$$

where w_C denotes the centralized (cooperative) solution. Clearly, the cooperative wage rate is bigger (smaller) than the noncooperative synchronized wage rate w_S (given by (28)) when the externalities existing between trade unions for the non-cooperative solutions are positive (negative); that is, as we have just seen, when $\lambda > \theta^{-1}$.

4. Comparison of the Solutions

We can now compare the solutions obtained in the three cases. We are interested in the wage rates and the levels of employment generated. The following proposition applies. The proof is given in the Appendix.

PROPOSITION 1. *Assuming that Lemma 1's conditions are fulfilled:*

(i) if $\theta > 1$ and $\lambda > \theta^{-1}$, then

$$w_C > w_I > w_S$$

and

$$l_C < l_I < l_S;$$

(ii) if $\theta > 1$ and $\lambda < \theta^{-1}$, then

$$w_S > w_I > w_C$$

and

$$l_S < l_I < l_C;$$

(iii) if $\theta < 1$ and $\lambda < \theta^{-1}$, then[7]

$$w_I > w_S > w_C$$

and

$$l_I < l_S < l_C.$$

According to this proposition, the relationship between the wage rate and the employment level depends on a simple condition between the two important parameters of the model: the openness index and the differentiation index, which determines the "market power" of national trade unions.

Proposition 1 states that, if both countries belonging to the MU are producing gross substitutable goods (θ bigger than 1) and are only slightly open to the rest of the world (λ small), then centralized bargaining will lead to smaller wage rates and, hence, higher employment. The opposite result obtains when the openness index and the differentiation index are high. Proposition 1 is easily understood when two characteristics of the economy, and more precisely, of the instantaneous payoff function, are taken into account: the nature of externalities and of strategic interactions.

Let us consider for example case (i). When λ is bigger than the inverse of θ, there exist positive externalities for the value of the wage rate obtained through decentralized and synchronized bargaining; as we have just seen, the positive output effect of a higher wage decided by agent j more than offsets the negative price effect on the wage rate in country i. When θ is bigger than 1, there exist strategic complementarities between the actions decided by both players. To understand the inequality between w_S and w_I, consider the case when decisions by national trade unions are staggered. Then, when trade union a must choose its wage rate to be valid for the two periods to come, it understands that, if it increases the wage rate in country a relative to its synchronized decentralized value, trade union b, next period, will increase the wage rate to prevail in country b, because of the existing strategic complementarities between both players. Since externalities are positive (the employment gain more than offsets the lesser increase in the price level), this has a net positive effect on the present value of the discounted future welfare of a. Hence, a is induced to choose a higher wage rate than the synchronized value w_S. Trade union b's reasoning similarly, when both trade unions use Markov strategies, the stationary solution corresponding to the Markov perfect equilibrium, w_I, is higher than w_C.

The inequality between w_C and w_S obtains immediately because of positive externalities.

Lastly, the intermediate position of w_I comes from the concavity of the instantaneous payoff function. Indeed the staggered time pattern allows each trade union to partially and noncooperatively take into account the existing externalities. This is due to the fact that every two periods, it is a temporary leader in the game: when it sets its decision, it knows the prevailing decision (the wage rate) in the other country. It is able to react to this precommitted wage rate and take into consideration the existing externalities between players. The other player is unable to renege and alter its current decision. It follows that the staggered decision is always closer to the cooperative solution than the noncooperative synchronized one.

The two other cases can be explained through similar reasoning, based on the signs of externalities and strategic interactions between players.

Endogenizing the Wage-Setting Time Pattern

Given that different wage-setting time patterns generate different outcomes, and in particular affect the level of employment achieved in the economy, it is worthwhile considering which time pattern is likely to emerge when trade unions behave noncooperatively. This problem of endogenizing the time pattern of decisions can be answered through a timing game,[8] being played by noncooperative trade unions. A timing game is played "beforehand," before the actual bargaining game on wages is taking place. The decision to be taken by each player is to choose to act (to set the wage rate in the present framework) in odd or in even periods. The outcome of such a game is the time pattern of decisions: if both players choose to act in odd (or in even) periods, decisions are synchronized; if one trade union plays in odd periods and the other one in even periods, decisions are staggered. To simplify the analysis and get rid of any transitional dynamics problem, we assume that, once the equilibrium of the timing game is determined, the stationary equilibrium of the wage-setting game is reached at the first period. Then, we are able to establish the following proposition, which is proved in the Appendix.

PROPOSITION 2. *In a noncooperative timing game, the trade unions will choose to stagger their wage bargaining decisions if $\theta > 1$ and to synchronize their wage bargaining decisions if $\theta < 1$.*

The timing game allows the trade unions to select noncooperatively the time pattern generating the highest level of employment when θ is less than one (as in this case, they choose to synchronize and $l_s > l_t$) or when θ is greater than one and $\lambda < \theta^{-1}$ (as in this case, they choose to alternate and $l_t > l_s$). But, by comparing Proposition 1 and Proposition 2, it appears that the best solution as far as employment is concerned will never obtain through a noncooperating behavior between the countries' trade unions, since a synchronized pattern of decision does not generate the highest employment level when $\theta < 1$, nor a staggered pattern when $\theta > 1$. Some institutional decision could then be welcome, aimed at modifying the pattern of bargaining within the MU.

Note that, in the case of a quasi-closed MU ($\lambda \to 0$), a "fortress union," a full integration of trade unions, would lead to a smaller wage rate and higher employment. The argument is often offered that European integration would be a means to constitute a "fortress Europe." That is, European integration has a protectionist objective and aims at forming a "fortress" able to erect tariff or nontariff barriers more successfully than weak and isolated European countries in order to protect its workers' standard of living from world competition, even at the cost of higher unemployment.

Without questioning the validity of this argument, we remark that a centralization of wage bargaining at the MU level, in such a case, would be a strong antidote since it would lead to lower wage rates and higher employment. This comes from the fact that the negative effects of a wage competition between trade unions within the MU are stronger the closer the MU is. Centralization of wage bargaining would allow internalization of the negative effects of competition on wages, and hence on prices. If such a pattern of bargaining were impossible to implement, staggered bargaining would be a "second-best" solution.

Endogenizing Monetary Policy

Until now we have assumed that money supply is set according to a constant rule, independent of the actual outcome of the economy. This could cast some doubts on the robustness of the results obtained. Wage-setting choices must account for monetary policy decisions, and vice versa, in a real-world context.[9] Indeed, when this assumption is relaxed and we endogenize in a simple way the behavior of the MU's central bank, our results are enriched but not qualitatively altered. To address the issue of the endogenizing of monetary policy, we make two simplifying assumptions. Firstly, the central bank adopts a discretionary behavior relative to the trade union(s). Secondly, the central bank sets the money supply for two periods.

The central bank's instantaneous payoff function is defined as follows:

$$G(y_a, y_b, p) = -\left\{\left[(y_a - \tilde{y})^2 + (y_b - \tilde{y})^2\right]\Big/2\right\} - gp^2/2, \quad g \geq 0. \tag{36}$$

The first term on the right relates to the output objective of the bank. Both countries are given the same weight in the central bank's payoff. For each country, the central bank has an output target $\tilde{y}$ and it wants the actual output to be as close as possible to this output level. Notice that this target level may differ from the trade unions' target level $\bar{y}$, reflecting the (possibly) differing viewpoints of the various protagonists upon unemployment and the welfare of the unemployed. The second term relates to inflation in the MU which the central bank would like to fully eradicate. The central bank's instrument is the money supply m^0 (log), which affects outputs and the MU's general price level through the exchange rate and the price of imports.

We consider the steady-state Nash equilibria of the three games we have previously defined, when the central bank and the trade union(s) behave noncooperatively. At the steady state of the Nash noncooperative decentralized synchronized game, the money supply is denoted $\hat{m}_S$ and the steady-state Nash solution is the four-tuple $(\hat{w}_S, \hat{l}_S, \hat{m}_S, \hat{p}_S)$. Similarly, the steadystate solution to the decentralized staggered game is $(\hat{w}_I, \hat{l}_I, \hat{m}_I, \hat{p}_I)$ and the steady-state solution to the centralized game is $(\hat{w}_C, \hat{l}_C, \hat{m}_C, \hat{p}_C)$. We can then generalize Proposition 1 and state the following (proved in the Appendix).

PROPOSITION 3. *Assuming that the conditions stated in the lemma are fulfilled:*

(i)   if $\theta > 1$ and $\lambda > \theta^{-1}$, then

$$\hat{w}_C > \hat{w}_I > \hat{w}_S \quad \hat{l}_C < \hat{l}_I < \hat{l}_S$$

and

$$\hat{m}_C > \hat{m}_I > \hat{m}_S \quad \hat{p}_C > \hat{p}_I > \hat{p}_S;$$

(ii) if $\theta > 1$ and $\lambda < \theta^{-1}$, then

$$\hat{w}_S > \hat{w}_I > \hat{w}_C \quad \hat{l}_S < \hat{l}_I < \hat{l}_C$$

and

$$\hat{m}_S > \hat{m}_I > \hat{m}_C \quad \hat{p}_S > \hat{p}_I > \hat{p}_C;$$

(iii) if $\theta < 1$ and $\lambda < \theta^{-1}$, then

$$\hat{w}_I > \hat{w}_S > \hat{w}_C \quad \hat{l}_I < \hat{l}_S < \hat{l}_C$$

and

$$\hat{m}_I > \hat{m}_S > \hat{m}_C \quad \hat{p}_I > \hat{p}_S > \hat{p}_C.$$

According to this proposition, the endogenizing of monetary policy does not modify the ranking of wages and employment levels obtained under a fixed money supply rule. Therefore, the various implications drawn from Proposition 1 remain valid. This comes directly from the assumption that the arguments of the objective functions of players are real variables only and there is no money illusion because of perfect information. Note that the ranking of money supplies and prices is identical to the ranking of wages in the three cases of the proposition. The central bank faces a tradeoff between output and inflation, for a given wage level. When the wage rate is lower, the MU becomes more competitive compared with the rest of the world, and this boosts production and employment. Therefore, the central bank becomes more concerned about inflation and sets the money supply at a lower level. This explains why, when for example $\theta > 1$ and $\lambda > \theta^{-1}$, money supply is lower when bargaining is both synchronized and decentralized.

These results are nothing more than insights into the difficult issue of interdependence between wage-setting choices and monetary policy when time patterns of decisions are taken into account. Two remarks are important. Firstly, the model is basically similar to the simple Barro–Gordon (1983) model: money is neutral. When there is no possibility to fool private agents, the central bank can never achieve a higher output at the expense of higher inflation; playing with discretion results only in higher inflation with no real gain. Logically this raises the issue of credibility and time-inconsistency. How to address this issue with Markov strategies is still unclear.

Secondly, our assumption that money supply is set for two periods is clearly open to question. Money supply obviously varies with greater frequency than wages set by trade unions. A more realistic assumption would generate a wealth of tricky problems, both economic and mathematical. For example, the fact that money supply can be modified more rapidly than wages gives the central bank a clear superiority over trade unions, which is not easily reconciled with the notion of Markov strategies. In brief, these remarks make it clear that the whole issue requires a more in-depth analysis which is beyond the scope of the present paper.

5. Conclusion

In an open monetary union, the issue of real adjustments inside the union is particularly important as economic integration between countries belonging to the monetary union widens. The aim of the present paper has been to evaluate the consequences on nominal remunerations and employment from alternative bargaining processes which differ according to their level and time pattern. Assuming that wage rates are fixed through bargaining for several periods, bargaining may (1) happen on the MU's level and be centralized, or on the "country" level and be decentralized; (2) be "synchronized" for all workers in the MU or be "alternate" (i.e., not take place at the same time in the various countries).

For simplicity we have looked at the case of a two-country MU, open to the rest of the world, where wages are precommitted for two periods. Wages can be set by national trade unions or by a central one, covering the whole MU. We supposed Markovian strategies for the players. Analyzing three specific bargaining structures, we showed that the knowledge of existing externalities and dynamic strategic interactions are crucial when comparing the three possible outcomes. Namely, from the point of view of employment, the superiority of one or the other of these processes depends on the openness index and the differentiation index between goods produced in the union.

When goods produced within the MU are gross complements, national trade unions prefer synchronized bargaining. When goods are gross substitutes, they prefer to alternate their decisions. But in any case, these patterns of bargaining are never the best with regard to employment. When goods are gross complements, bargaining should take place at the MU's level; i.e., be centralized. When goods are gross substitutes, two cases arise. If the openness index towards the rest of the world is small compared with the inverse of the differentiation index between goods, it is preferable that bargaining takes place at the MU's level and a central trade union should improve the employment prospect: a "corporatist" MU would be justified because (the externalities within the union being negative) a centralized bargaining would lead to a lower wage rate and allow the MU to be more competitive relative to the rest of the world. In the opposite case, it is preferable to set synchronized but decentralized bargaining. The endogenizing of discretionary monetary policy does not modify these conclusions.

Important issues have been left aside in the present paper, such as uncertainty and more elaborate behavior by the MU's central bank. To enlarge the present framework by taking these features into consideration is left to further research.

Appendix

The Existence of a Stable Unique Markov Perfect Equilibrium of the Staggered Game

The first-order condition corresponding to program (32) is

$$v_1\left[R^{il}(w_i), w_j\right] + \delta W_i'\left(R^{il}(w_j)\right) = 0. \tag{A1}$$

The function $W_i(w_i)$ can be defined as

$$W_i(w_i) = v\left[w_i, R^{il}(w_i)\right] + \delta v\left[R^{il}\left(R^{il}(w_i)\right), R^{il}(w_i)\right] + \delta^2 W_i\left[R^{il}\left(R^{il}(w_i)\right)\right]. \tag{A2}$$

Given that by assumption $R^{il}(w_j) = \alpha_i + \beta_i w_j$, $W_i''(w_i)$ is constant, and

$$W_i'' = \left[1 - \delta^2 \beta_i^2 \beta_j^2\right]^{-1}\left[v_{11} + 2v_{12}\left(1 + \delta\beta_j\right)\beta_i\beta_j\right.$$
$$\left. + \left(1 + \delta\beta_j v_{22}\beta_j^2\right) + \left(\delta v_{11} + 2v_{22}\right)\beta_i^2\beta_j^2\right]. \tag{A3}$$

Differentiating (A1), one obtains

$$\partial R^{il}(w_i)/\partial w_i \equiv \beta_i = -v_{12}/\left(v_{11} + \delta W_i''\right). \tag{A4}$$

Given (A2) and (A3), it is deduced from (A4) that

$$\beta_i v_{11}\left(1 + \delta\right) + 2\beta_i\beta_j\delta v_{12} + \beta_i^2\beta_j\delta\left(1 + \delta\right)v_{22} + \delta^2 v_{12}\beta_i^2\beta_j^2 + v_{12} = 0. \tag{A5}$$

Equation (A5) defines two equations for β_1 and β_2. Substracting one equation from the other yields

$$\left(\beta_1 - \beta_2\right)\left(v_{11} + \delta\beta_1\beta_2 v_{22}\right) = 0. \tag{A6}$$

Equation (A6) can have both symmetric and asymmetric solutions. An asymmetric solution is such that $\beta_1\beta_2 = -v_{11}/\delta v_{22}$. It is unstable if $|v_{11}| > |\delta v_{22}|$. Since $|v_{11}| = ((1 + \theta)/2)^2$ and $|v_{22}| = ((\theta - 1)/2)^2$, this condition is always fulfilled, $\forall \theta, \theta \neq 1$. Hence, an asymmetric solution can be ruled out. A symmetric solution implies that β is a root of $P(\beta)$:

$$P(\beta) = \delta^2 v_{12}\beta^4 + \delta\left(1 + \delta\right)v_{22}\beta^3 + 2\delta v_{12}\beta^2 + \left(1 + \delta\right)v_{11}\beta + v_{12}. \tag{A7}$$

Therefore, there is a unique stable solution to the staggered game if $P(\beta)$ has a unique real root inferior to one in absolute value. Given the instantaneous payoff function, $P(\beta)$ is equal to

$$P(\beta) = -4^{-1}\left[\left(\delta\left(1 - \theta^2\right)\beta^2 + \left(1 + \delta\right)\left(1 - \theta\right)^2\beta + 2\left(1 - \delta^2\right)\right)\delta\beta^2\right.$$
$$\left. + \left(1 + \delta\right)\left(1 + \theta\right)^2\beta + \left(1 - \theta^2\right)\right] \tag{A8}$$

Note that

$$P(1) = -(1+\delta)/4[2\delta+2+2\theta(1-\delta)] < 0,$$
$$P(-1) = (1+\delta)\delta[\delta\theta+\theta+1-\delta]/2 > 0.$$

Thus there are either one or three roots in the interval $]-1,1[$. Let us define sufficient conditions to obtain only one root within this interval. We deduce from $P(\beta)$ that

$$P'(\beta) = -4^{-1}\Big[4\delta^2(1-\theta^2)\beta^3 + 3\delta(1-\delta)(1-\theta)^2\beta^2 + 4\delta(1-\theta^2)\beta + (1+\delta)(1+\theta)^2\Big],$$
$$P''(\beta) = 3\delta^2(\theta^2-1)\Big[\beta^2 - (1+\delta)(1-\theta)/2\delta(\theta+1) + 1/3\delta\Big].$$

There are two real roots corresponding to $P(\beta)$ and hence only one root inferior to one in absolute value if $P(\beta)$ has no inflexion point; i.e., if polynomial $P''(\beta)$ does not accept any real root. This is true if

$$g(\delta) \equiv \delta/(1+\delta)^2 > 3/16 \cdot (\theta-1)^2/(\theta+1)^2 \equiv h(\theta).$$

Function $g(\delta)$ is increasing with δ over the interval $]0,1[$. Function $h(\theta)$ decreases from $3/16$ to 0 when θ goes from 0 to 1, and increases from 0 to $3/16$ when θ goes from 1 to infinity. Hence a sufficient condition for $P(\beta)$ to have a unique root belonging to $]-1,1[$ is that $\delta/(1+\delta)^2$ be bigger than $3/16$. When θ is bigger than 1, this root is positive since $P(0)$ is positive. When θ is smaller than one, this root is negative since $P(0)$ is negative.

Comparison of the Solutions

At steady-state equilibrium, where $w_i = w_j = w_l$, the derivative of $W_i(w_l)$, defined in (A2), yields

$$W_i(w_l) = \Big[v_1(w_l, w_l)(1+\beta\delta^2) + v_2(w_l, w_l)(1+\delta)\beta\Big]/(1-\beta^2\delta^2). \tag{A9}$$

Combining (A7) and (A4) yields

$$v_1(w_l, w_l) = v_{12}/(v_{11}+\delta W_i'')\delta v_2(w_l, w_l) = -\beta\delta v_2(w_l, w_l). \tag{A10}$$

First, we have to compare w_l, defined by (A3) and (A10), with w_S, defined by $v_1(w_S, w_S) = 0$ (according to (26)).

Note that $v_2(w, w)$ is linear and can be either increasing or decreasing, depending on the value of θ:

$$\partial v_2(w, w)/\partial w \equiv v_{12} + v_{22} = (\theta-1)/2 > 0 \Longleftrightarrow \theta > 1.$$

Note also that $v_1(w, w)$ is linear and always decreasing. The derivative $v_2(w_i, w_j)$ is equal to

$$v_2(w_i, w_j) = -(\theta-1)/2[\overline{m}-\overline{y}-(\theta+1)w_i/2+(\theta-1)w_j/2]-\gamma(1-\lambda)/2,$$

and at the synchronized equilibrium

$$v_2(w_S, w_S) = (\theta-1)[\overline{m}-\overline{y}-w_S]/2-\gamma(1-\lambda)/2.$$

Hence:

$$v_2(w_S, w_S) > 0 \Longleftrightarrow (\theta-1)(1+\lambda) > (\theta+1)(1-\lambda) \Longleftrightarrow \lambda > \theta^{-1}.$$

Then three cases must be distinguished:

(i) $\theta > 1$ and $\lambda > \theta^{-1}$
In this case, v_{12} is positive and $-\beta\delta v_2(w, w)$ is decreasing (since β is positive and $v_2(w, w)$ is increasing). The slope of $v_1 (w, w)$ is larger in absolute value than the slope of $v_2(w, w)$ since $1 + \theta > \theta - 1$. Then, as $\lambda > \theta^{-1}$ and therefore $v_2(w_S, w_S)$ is positive, this implies that $-\beta\delta v_2(w_S, w_S) < 0$, and therefore $w_I > w_S$.

(ii) $\theta > 1$ and $\lambda > \theta^{-1}$
As before, the slope of $v_1(w, w)$ is larger in absolute value than the slope of $v_2(w, w)$. But, now $v_2(w_S, w_S)$ is negative and this implies that $-\beta\delta v_2(w_S, w_S) > 0$, and therefore $w_I < w_S$.

(iii) $\theta < 1$ and $\lambda < \theta^{-1}$
In this case, v_{12} is negative and $-\beta\delta v_2(w, w)$ is decreasing (since β is negative and $v_2(w, w)$ is decreasing). The slope of $v_1(w, w)$ is larger in absolute value than the slope of $v_2(w, w)$ since $1 + \theta > 1 - \theta$. Then, as $\lambda < \theta^{-1}$ and therefore $v_2(w_S, w_S)$ is negative, this implies that $-\beta\delta v_2(w_S, w_S) < 0$, and therefore $w_I > w_S$.

From the concavity of function $v(w, w)$ and the sign of externality at $v(w_S, w_S)$, Proposition 2's results on wage levels obtain.

In order to obtain the inequalities on employment levels, we may note that, for each country, at the stationary equilibrium, the demand for its good is equal to

$$l = y = \overline{m} - p = \overline{m} - w.$$

Hence $\partial y/\partial w > 0$. Proposition 2's inequalities on employment levels follow.

The Timing Game

Consider both trade unions noncooperatively choosing their decision periods, "odd" or "even". If both play odd or even, they get $v(w_S, w_S)$; if one plays odd and the other one plays even, they get $v(w_I, w_I)$.

(i) $\lambda > \theta^{-1}$ and $\theta > 1$
We know from Proposition 2 that $w_C > w_I > w_S$. As $v(w, w)$ is concave with positive spillovers at w_S, we deduce that

$$v\!\left(w_C, w_C\right) > v\!\left(w_I, w_I\right) > v\!\left(w_S, w_S\right).$$

(ii) $\lambda < \theta^{-1}$ and $\theta > 1$
We know from Proposition 2 that $w_C < w_I < w_S$. As $v(w, w)$ is concave with negative spillovers at w_S, we deduce that

$$v\!\left(w_C, w_C\right) > v\!\left(w_I, w_I\right) > v\!\left(w_S, w_S\right).$$

(iii) $\lambda < \theta^{-1}$ and $\theta < 1$
We know from Proposition 2 that $w_C > w_I > w_S$. As $v(w, w)$ is concave with negative spillovers at w_S, we deduce that

$$v\!\left(w_C, w_C\right) > v\!\left(w_S, w_S\right) > v\!\left(w_I, w_I\right).$$

Therefore, if there is strategic complementarity ($\theta > 1$), the trade unions's welfare is higher when they set wages sequentially. If there is strategic substitutability ($\theta < 1$), the trade unions's welfare is higher when they set wages simultaneously. Therefore, "odd–odd" or "even–even" is a Nash equilibrium to the timing game if and only if there is strategic substitutability.

Monetary Policy

The trade union's payoff function has only real arguments. Money neutrality prevails and the ranking of wages among the three outcomes is identical to the ranking of money supplies. Remark then that, from (1):

$$p_x = e + p_x^* = m^0 + \varphi,$$

where $\varphi = p^* + \log(\alpha\lambda/(1-\alpha)D^*)$. Hence p is an increasing function of m^0. Using (20), (22), and (36), the central bank's instantaneous payoff function is equal to

$$G(y_a, y_b, p) = -2^{-1}\left[\left(m^0 + (\theta-1)\theta^{-1}w_a + (\theta+1)w_b/2 - \tilde{y}\right)^2\right.$$
$$+ \left(m^0 + (\theta-1)\theta^{-1}w_b + (\theta+1)w_a - \tilde{y}\right)^2 /2\bigg]$$
$$-2^{-1}g\left[(1-\lambda)(w_a+w_b)/2 + \lambda(m^0+\varphi)\right]^2.$$

Then, using (3) and deriving it with respect to m^0, we get

$$p = -(g\lambda)^{-1}\left[(y_a - \tilde{y}) + (y_b - \tilde{y})\right].$$

In steady-state equilibrion, we therefore get

$$\hat{p}_Z = -(2g\lambda)^{-1}(\hat{y}_Z - \tilde{y}), \quad Z = S, I, C,$$

from which Proposition 4 is easily deduced.

References

Ball, Laurence and Stephen Cecchetti, "Wage Indexation and Discretionary Monetary Policy," *American Economic Review* 81 (1991):1310–19.

Barro, Robert J. and David B. Gordon, "Rules, Discretion and Reputation in a Model of Monetary Policy," *Journal of Monetary Economics* 12 (1983):101–21.

Blanchard, Olivier J. and Stanley Fischer, *Lectures on Macroeconomics*, Cambridge, MA: MIT Press, 1989.

Bleaney, Michael, "Central Bank Independence, Wage Bargaining Structure and Macroeconomic Performance in OECD Economies," *Oxford Economic Papers* 48 (1996):20–38.

Cahuc, Pierre and Hubert Kempf, "Alternative Time Patterns of Decisions and Dynamic Strategic Interactions," working paper 95-11, Department of Economics, European University Institute, Florence, 1995, forthcoming in the *Economic Journal*, 1997.

Cahuc, Pierre and André Zylberberg, "Niveaux de Négociations Salariales et Performances Macroéconomiques," *Annales d'Économie et de Statistiques* 23 (1991):1–12.

Calmfors, Lars, "Centralization of Wage Bargaining and Macroeconomic Performance: A Survey," seminar paper 536, Institute for International Economics, Stockholm University, 1993.

Calmfors, Lars and John Driffill, "Bargaining Structure, Corporatism and Macroeconomic Performance," *Economic Policy* 6 (1988):12–61.

Canzoneri, Matthew B. and Dale Henderson, *Monetary Policy in Interdependent Economies*, Cambridge, MA: MIT Press, 1991.

Cooper, Russell and John Haltiwanger, "Macroeconomic Implications of Production Bunching: Factor Demand Linkages," *Journal of Monetary Economics* 30 (1992):107–27.

Cooper, Russell and Andrew John, "Coordination Coordination Failures in Keynesian Models," *Quarterly Journal of Economics*, 103 (1988):441–63.

Cubbit, Robin P., "Monetary Policy Games and Private Sector Commitments," *Oxford Economic Papers* 44 (1992):513–30.

De Fraja, Giovanni, "Staggered vs Synchronised Wage Setting in a Duopoly," *European Economic Review* 37 (1993):1057–69.

Dixon, Huw, "Macroeconomic Policy in a Large Unionised Economy," *European Economic Review* 35 (1991):1427–48.

——, "Imperfect Competition and Open Economy Macroeconomics," in Frederick van der Ploeg (ed.), *Handbook of International Macroeconomics*, Oxford: Basil Blackwell, 1994:31–61.

Dixon, Huw and Neil Rankin, "Imperfect Competition and Macroeconomics," *Oxford Economic Papers* 46 (1994):171–99.

Driffill, John and Frederick van der Ploeg, "Monopoly Unions and the Liberalisation of International Trade," *Economic Journal* 103 (1993):379–85.

Fethke, Gary and Andrew Policano, "Wage Contingencies, the Pattern of Negotiation and Aggregate Implications of Alternative Contract Structure," *Journal of Monetary Economics* 14 (1984):151–70.

——, "Will Wage Setters Ever Stagger Decision?" *Quarterly Journal of Economics* 101 (1986):867–77.

——, "Monetary Policy and the Timing of Wage Negotiations," *Journal of Monetary Economics* 19 (1987):89–105.

——, "Information Incentives and Contract Timing Patterns," *International Economic Review* 31 (1990):651–65.

Fischer, Stanley, "Long Term Contracts, Rational Expectations and the Optimal Money Supply," *Journal of Political Economy* 85 (1977):191–206.

Kempf, Hubert, "Inflation and Wage Indexation with Multiperiod Contracts: A Comment," *European Economic Review* 33 (1989):1397–404.

Maskin, Eric and Jean Tirole, "A Theory of Dynamic Oligopoly. I: Overview and Quantity Competition with Large Fixed Costs," *Econometrica* 56 (1988a):549–69.

——, "A Theory of Dynamic Oligopoly. II: Price Competition, Kinked Demand Curves and Edgeworth Cycles," *Econometrica* 56 (1988b):570–99.

——, "A Theory of Dynamic Oligopoly. III: Cournot Competition," *European Economic Review* 31 (1989):947–96.

Rasmussen, Bo S., "Exchange-Rate Policy, Union Wage Indexation and Credibility," *Journal of International Economics* 35 (1993):151–67.

Spivak, Avia, J. Weinblatt, and B. Z. Zilberfarb, "Inflation and Wage Indexation with Multiperiod Contracts," *European Economic Review* 31 (1987):1299–312.

Tabellini, Guido, "Centralized Wage Setting and Monetary Policy in a Reputational Equilibrium," *Journal of Money, Credit and Banking* 20 (1987):102–18.

Waller, Christopher and David VanHoose, "Discretionary Policy and Socially Efficient Wage Indexation," *Quarterly Journal of Economics* 107 (1992):1451–60.

Notes

1. Throughout the paper, "monetary union" (MU) will designate this two-country partnership, sharing the same currency with free trade. The employees' bargaining units will be designated as "trade unions."

2. The macroeconomics of imperfect competition has made impressive progress in recent years (see the survey by Dixon and Rankin, 1994), although its extension to the case of an open economy is much less developed (see the survey by Dixon, 1994).

3. In a series of papers (1984, 1986, 1987, 1990), Fethke and Policano studied various aspects of wage bargaining timing in a closed economy with both sectorial and aggregate shocks. The closer to the present issue is their first paper, in which they show that the staggering of wage contracts is welfare-improving when idiosyncratic shocks are important relative to aggregate ones, and that wage indexation enhances the benefits of staggered negotiations. However, their solution concept differs from ours and fails to take into consideration strategically the dynamic interactions between sectors.

4. Here, profits are nil because of constant returns to scale and perfect competition.

5. This assumption can be justified by means of a refinement in the structure of the world economy. Let us now consider that the world economy consists of countries of two types. On the one hand, "North" countries produce and export to "South" countries two goods, A and B. These two goods are imperfect substitutes. The monetary union we consider belongs to the North. It is formed of two countries. Each country belonging to the MU is specialized in the production of one good: country A produces good A and country B produces good B. Altogether, the MU is producing and exporting a composite good "U". On the other hand, "South" countries produce and export to "North"countries a good "X". There are N "North" countries in addition to the MU and S "South" countries. Both numbers are large. For convenience and without loss of generality, we consider a single "South" currency and a single exchange rate between the MU and the "South" countries. The MU's current account balance is always cleared.

Consumer preferences in the MU are those given in the text. Consumers of the MU want to consume only goods A and B produced in the MU and good X produced in "South" countries. Consumers in "South" countries want to consume all goods produced by "North" countries. For instance, the "South" consumer preferences are assumed to be given by the following expression:

$$U^S\left(C_{MU}, C_1, \ldots, C_i, \ldots, C_N, C_X, M/P\right) = C_{MU}^{\alpha(1-\lambda)\mu} \cdot \prod_{i=1}^{N} C_i^{\alpha(1-\lambda)h_i} \cdot C_X^{\alpha\lambda} \cdot (M/P)^{1-\alpha}$$

$$0<\alpha<1, \quad 0<\lambda<1, \quad \mu+\sum_{i=1}^{N}h_i = 1,$$

where C_{MU} is the consumption of good U produced by the MU, C_i is the consumption of (composite) good produced by "North" country i, C_X is the consumption of good X. "South" preferences over good U are given by (7).

It is assumed that each "North" country having a negligible weight on world markets receives a constant fraction of total demand from "South" countries. The global import of goods A and B by "South" countries corresponds to a constant share of their total expenditures, thanks to consumer preferences we use.

6. This approximation is used by Blanchard and Fischer (1989). From (8), we obtain $(P_u/P_a)^{1-\theta}$ $= [1 + (P_b/P_a)^{1-\theta}]/2$. In the neighborhood of the symmetric equilibrium, $P_b \to P_a$ and $P_u \to P_a$. Therefore:

$$\left(P_u/P_a\right)^{1-\theta} \to \log\left(P_u/P_a\right)^{1-\theta} +1,$$
$$\left(P_b/P_a\right)^{1-\theta} \to \log\left(P_b/P_a\right)^{1-\theta} +1.$$

This implies
$$(1-\theta)p_u - (1-\theta)p_a + 1 = \left[2 + (1-\theta)p_b - (1-\theta)p_a\right]/2;$$
hence the approximation written in the text.

7. The case $\theta < 1$ and $\lambda > \theta^{-1}$ is impossible since $\lambda < 1$.

8. We adopt the same timing game as Cooper and Haltiwanger (1992).

9. A growing literature is devoted to this issue. In closed economies, see for example Ball and Cecchetti (1991), Bleaney (1996), Cubbitt (1992), Dixon (1991), Tabellini (1987), Waller and VanHoose (1992), and in a small open economy Rasmussen (1993).

Review of International Economics, Special Supplement, 111–133, 1997

The Benefits of Environmental Fiscal Reforms in an Integrated Europe

*Carlo Carraro and Marzio Galeotti**

Abstract

This paper analyses the effects of an environmental fiscal reform where the environmental tax revenue is used to reduce the tax burden on labor in order to increase employment. The study tries to assess the benefits generated by this type of reform in an integrated Europe where different degrees of tax harmonization can be achieved. The differential impact of cooperative fiscal reforms versus noncooperative ones is quantified through the use of a newly developed general equilibrium model. The results suggest that only a high degree of harmonization, in which both revenue and expenditure policies are coordinated, can provide significant environmental and employment benefits.

1. Introduction

There are at least three important issues which have been debated in Europe within the discussion on the gains and losses associated with economic integration among European Union (EU) countries. The first issue is a traditional one and concerns the macroeconomic gains from international coordination. Coordination may be desirable because uncoordinated policymaking may ignore the consequences of spillovers from one country to another, so that policy in the aggregate will result in unsatisfactory outcomes. This issue has recently been analyzed in the context of monetary policy (Hughes-Hallett and Ma, 1985; Mélitz and Weber, 1996), budgetary policy (Hughes-Hallett and McAdam, 1996), and trade policy (Veugelers and Vandenbussche, 1996).[1] Within this context, however, a few analyses on the benefits from the international coordination of environmental policies have been carried out.[2] Hence the need for a careful assessment of which are the effects of environmental policymaking in an integrated Europe where policy decisions need to be harmonized.

A second issue concerns the impact of environmental policies where the revenue raised by an emission tax is recycled either to foster economic growth (through appropriate expenditure policies) or to reduce existing distortions in the tax system. In particular, among the possible uses of the environmental tax revenue the one in which it is redistributed in the form of a general reduction in social security payments by employers has received the greatest attention (Drèze et al., 1993). The main goal of this type of reform, which shifts the tax burden away from labor toward environmental and natural resource inputs (chiefly, energy) is to achieve the so-called "employment double dividend," which is a decrease of both emissions and unemployment (Carraro et al., 1996).

The recent theoretical literature on the "employment double dividend" hypothesis has been unable to reach clear-cut conclusions.[3] Nonetheless, it seems to indicate that very restrictive conditions must be met for the recycling of environmental taxation to generate an increase in aggregate employment (Bovenberg, 1997). Moreover, even when these conditions are met, the effects on employment of a fiscal reform which

---

*Carraro: University of Venice, Ca' Foscari, 30123 Venice, Italy. Email: ccarraro@unive.it. Galeotti: University of Bergamo, Bergamo, Italy. Email: galeotti@unibg.it.

reduces distortionary taxation in the labor market and increases environmental taxation are likely to be small (Bovenberg and Goulder, 1993). The theoretical results are largely supported by the evidence of empirical analyses, even if the latter show a large heterogeneity of conclusions.[4] It is noticeable that theoretical results are usually obtained either in the case of a closed economy or in the case of an open economy in which only an individual country introduces the fiscal reform. On the other hand, empirical contributions sometimes discuss the introduction of the fiscal reform in several countries (Denis and Koopman, 1994; Capros et al., 1996; Carraro et al., 1996), but do not discuss the additional gains (or losses) that can be achieved by coordinating the fiscal reform.[5]

A third important issue is the institutional design of policymaking in an integrated Europe. There has been a debate on the power to be assigned to the European Commission and the European Parliament, on the role of the future European Central Bank, on the policies that will be set at the national level. More specifically, the subsidiarity principle has been advocated as a discerning criterion which can help understanding in which cases decisions have to be taken by national governments. The issue of the institutional design is probably crucial also for environmental fiscal reforms and deserves a careful analysis. The questions to be answered concern generally which kind of institutional framework is the most appropriate to introduce an emission tax with revenue recycling aimed at boosting employment. More precisely, the problem is whether the environmental fiscal reform should be designed and implemented at the central level by the European Commission, or should be decentralized according to the subsidiarity principle. The latter option seems to receive a large consensus among policymakers. The difficulty to move toward a European carbon tax is probably the explanation of this change of direction which allows each country to set its own policy mix. However, is really subsidiarity the best institutional framework within which an environmental fiscal reform can be implemented?

This paper is a first attempt to tackle the above issues. We compare the effects of an environmental fiscal reform designed to reduce emissions and to increase employment carried out under different degrees of international coordination. The first option which we are going to assess is that of an emission tax harmonized across EU countries (thereby equalizing marginal abatement costs), but where the revenue is left to each national government which decides how to spend it to reduce payroll taxes without coordinating this subsidy to employers with the other EU partners. The second possibility is the one in which the subsidiarity principle is invoked. Countries agree upon the general structure of the reform to be implemented, but autonomously set the tax rates. This implies that each country decides the level of its own tax rate on carbon dioxide emissions and uses the fiscal revenue to reduce social security contributions paid by employers. The third policy examined achieves a high degree of harmonization because both environmental taxation and labor recycling policies are coordinated. Thus, emission tax rates equalize abatement costs, and the fiscal revenue is used by the European Commission to reduce European unemployment; that is labor costs are reduced more where unemployment rates are highest.

It is worth observing that when the decision on the design of the environmental fiscal reform is taken at the federal level two gains should, at least in principle, be achieved. First, free-riding behavior in some countries ought to be avoided; second, an explicit system of transfers in favor of some member states may take place, thus increasing the profitability of the fiscal reform at the EU level. This paper aims to provide some preliminary information on the actual gains following from coordinating decisions as far as an environmental fiscal reform is concerned.

In order to carry out the above analysis a new quantitative model of the European Union which is designed to assess the impact of environmental policies on growth, employment, and innovation activity is used. Being more general and disaggregated than theoretical models, and more accurately designed than previous European econometric models, it should enable us to obtain a more reliable empirical evaluation of the "employment double dividend" and related issues.[6] In particular, the model—called WARM (World Assessment of Resource Management)—accomplishes two main tasks: it captures the relevant features of the European labor market by explicitly modeling the bargaining process between unions and firms, and it describes the behavior of the European economies in an integrated framework in which each member country is viewed as a region belonging to the EU economic system.

In section 2 the main features of the WARM model are briefly described. Section 3 is devoted to the presentation of the main results of our simulation experiments. The evidence suggests that the gains from international coordination are not large, a finding which is common also to the literature on monetary cooperation (Carraro and Giavazzi, 1987). However, when a high degree of harmonization in which both revenue and expenditure policies are coordinated can be achieved, the environmental fiscal reform provides significant environmental and employment benefits. Our attempt to draw a few policy guidelines and suggest directions for future research conclude the paper.

2. The Econometric General Equilibrium Model

WARM represents the behavior of the production, household, government, and foreign sectors. In this section we provide a cursory description of the model. A full description can be found in Carraro and Galeotti (1996), to which the reader is referred for many more details.[7] The WARM model shares many features of top-down econometric general equilibrium models of which a prominent example is that of Jorgenson and Wilcoxen (1990). In particular, it describes the behavior of economic agents by means of demand and supply relationships which clear the markets for goods and services. Moreover, it makes use of flexible functional forms to parametrize tastes and technologies. The novel features of WARM can be summarized as follows.

(1) The model describes the economic structure of 12 European countries as well as that of the (pre-1995) European Union (EU12 hereafter) as a whole. Rather than adopting an interlinked methodology, it integrates the country-specific differences from a common European denominator within a unified and homogeneously designed framework. Estimation-wise, all the country observations for each variable are put together to form a panel dataset for which appropriate econometric techniques are used.

(2) The model devotes special attention to the issue of technical change. Its effects are incorporated in the firm's decision rules via an index of the quality of capital stock. This index is constructed using a latent variable approach where a Kalman filter technique permits the use of some relevant economic indicators (including R&D expenditures, patent imports, GDP growth) to decompose the time series of the capital stock into two components, the "environment-friendly" one and the "polluting" one, which then yield an index of the quality of the capital stock.

(3) Markets are not assumed to be perfectly competitive. In particular, the labor market is segmented. In the primary labor market, pertaining to the leading production sector, a trade union is active and plays a Nash bargaining game with

the employer's representatives that yields the equilibrium wage structure, which in turns affects the employment–wage relationships of the whole economic system. This feature is especially relevant for the European countries.

The Production Sector

Given the main intended uses of the model, special efforts are made in describing the production, consumption, and exchange of energy sources. In order to properly analyze the economic implications of CO_2 emissions, the WARM model allows for substitution possibilities among fossil fuels, substitution among fossil and nonfossil fuels, substitution among energy and other inputs and, finally, product substitution in the consumption mix. The WARM model further allows for the multiplicity of economic activities (agriculture, industry, transportation, energy production) which generate emissions to different degrees and account for the other greenhouse gases, besides carbon dioxide.

To accomplish the above tasks, we distinguish four types of production activities carried out by representative firms operating in each European country. The manufacturing firm is involved in the production and distribution of three nonagricultural nonenergy outputs consisting of durables, nondurables, and services.[8] The factors used are the services from labor, capital, energy, and a nonenergy intermediate input. The demand for this intermediate good results from the aggregation of agricultural products and imported (durable and nondurable) goods and services. The energy aggregate is made up of electricity and fossil fuels. These fuels in turn are comprised of coal, gas, and petroleum products (divided into products for industrial use and for transportation). Electricity is supplied by the electricity firm which uses capital, labor, and energy sources as inputs. This firm also directly imports electricity from abroad to be distributed nationwide. The fossil fuel transformation firm (or energy firm for short) supplies and distributes nonelectric energy sources to all other firms and to households combining primary factors along with imported fuels. Finally, the agricultural output is entirely supplied to the manufacturing sector by an agricultural firm which combines capital, labor, land, and energy.

We decompose the set of decisions made by firms into three different conceptual stages. First, given the amounts of outputs and capital stock, the cost-minimizing levels of variable inputs are determined. Through appropriate maintained assumptions about the technology, this problem is formulated as a multistage decision process and yields sets of interrelated input demands. Next, variable profits are maximized by selecting the optimal output levels: this process generates a set of inverse supply relationships or price functions. Finally, the optimal investment decisions are taken by the firm in such a way that total long-run profits are maximized.

The technological constraint is embodied in the firm's variable cost function which is parametrized as a modified generalized Leontief flexible functional form (Morrison, 1988). Application of Shephard's Lemma yields the system of demand functions for variable inputs. The firm's production structure is assumed to be homothetically separable technology, so that relevant simplifications in the design of the input decision process can be achieved: these choices proceed along several stages where the optimal mix within homogeneous groups of inputs is found and the aggregate input quantities, to be used in higher stages of the decision tree, are formed. On the output supply side, given the assumed generalized Leontief technology, a marginal cost function for each output produced can be computed. To obtain the firm's supply relationships, we assume that in general imperfect competition prevails.[9] To describe the price formation process, it is assumed that firms are price-makers and apply a mark-up pricing rule. In the current version of the model the price mark-up is assumed to be constant and time-invariant. At a second stage, producers supply their outputs to different customers according to a price discriminating strategy. Given the price function for each type of output, the price charged to each customer is chained to the corresponding output price by means of a proportionality factor. Note that the joint estimation in the model of these price functions together with the corresponding output demand relationships originating from the other sectors ensure that market-clearing conditions are enforced.[10]

To complete the description of the production side, the firm's investment decision is based on the dynamic comparison between the actual and the desired endowment of capital stock. As is well known, given the structure of the technology, the desired amount of capital input is determined from the variable cost function via the envelope condition. In the long-run equilibrium, the user cost of capital must be equal to its shadow value, given by the savings in marginal variable costs allowed by an extra unit of capital.[11]

The Household Sector

The basic behavioral assumption that characterizes a household's decisions is the maximization of the lifecycle utility subject to a budget constraint that equates expenditures to available resources. These are given by labor income, profits distributed by firms, and the return on financial wealth. In the model, financial wealth takes essentially the form of government bonds. The utility of a household is a function of leisure and overall consumption: however, for convenience, it is assumed that the utility function is separable in the two arguments, such that consumption choices can be treated independently from labor supply decisions.[12]

Consumption decisions take place in a sequential way. First, resources are intertemporally allocated, so that current and future consumption (savings) are determined. At a second stage, intratemporal consumption decisions are taken. In a manner analogous to the production sector, this is a multistage decision process, each generating systems of consumer demands for homogeneous groups of goods and services. Following the utility tree, the household selects the consumption of energy sources, other nonduable goods, services from durables, and other services. The durables decision is predetermined relative to the other choices and is described below. It is assumed that each allocation stage is described by a rationed "almost ideal demand system" (AIDS). The explanatory variables are the price of the jth good, the technical change indicator previously described, the number of households in a given country, and the stock of durable goods and of residential buildings. Exploiting the maintained assumption of weak separability of consumer preferences, and using again the AIDS approach, the distribution of total spending on energy between transportation and residential purposes (heating and the like) is obtained. Moreover, expenditures on energy for residential use are also allocated among several energy sources.

As for the acquisition of durable goods, in a manner analogous to a firm's investment, households choose the level of the desired stock and dynamically adjust the existing stock to this target. The desired amount of durables is determined from the indirect utility function generating the AIDS system. The stock of durables affects the demand of all goods, including energy.

Technological Change

One of the salient features of the model is an explicit attempt at describing the mechanism through which economic variables affect technological change. The basic idea is that the dynamics of technical change cannot be observed. Rather than following the traditional approach, according to which technical change is proxied either by a time trend or by the computation of factor productivity indexes, we adopt a latent variable structural equation approach. (See the Appendix for a detailed presentation of our model of endogenous technical change.) This procedure extracts information from indicators and cause variables, such as total expenditures on public and private R&D, imports of patents, and business cycle indicators, while being able to avoid using them as exact representations of technological change.[13]

It is assumed that the capital stock is composed by two components: an energy-saving stock and an energy-consuming one. Each year a new vintage of capital becomes operational. In this way new capital is added to the two components. The characteristics of this new capital depend on a number of economic variables which affect the firm's decision with regard to installing energy-saving capital (e.g., the energy prices). More precisely, relative prices, including the price of energy, affect both the decision to carry out R&D and the composition of the capital stock (e.g., higher energy prices induce firms to invest in energy-saving vintages of the capital stock). In the model, energy-saving technical progress is measured by the ratio between the two components of the capital stock. From our estimates, the average growth rate of this indicator is fairly low in the developed EU countries (about 2% in Germany, France, and Italy, and slightly lower in the UK), whereas it is much higher in the less developed countries (from 9% in Ireland to 30% in Greece). In all countries, the growth rate of the technical progress indicator becomes lower as the country grows (because the model objective is to capture the implementation of best-available technologies in the short- and medium-run).

The dynamics of our technical progress indicator generally affects the decision rules of all the agents in the model: not only does it affect the firm's input and output decisions, but it also influences a household's choices, especially those pertaining to the consumption of energy sources. Moreover, the dynamics of technical change concerns total private capital stock, and not only the component used as a production input by firms. An important implication of our treatment of environmental technical change is that it is a diffuse phenomenon. In fact, the technological indicator is an argument of the behavioral equations concerning the demand for inputs and the supply of outputs (via marginal costs). In particular, this is the case for the manufacturing firm. Thus, over time an increasing amount of environment-friendly capital is used in production which translates into an improvement of the production process and in turn of the quality of goods and services supplied. To the extent that these goods and services are purchased by the other firms and households, the process of endogenous technical change affects all sectors of the economy (and the "rest of the world" as well).[14]

Labor Markets, Unions, and the Wage Rate

The labor market in the model is not competitive but segmented. In particular, we distinguish between a primary sector and a residual sector. The former regards all production activities including the public sector but excluding agriculture; wages and employment in it are the outcomes of a bargaining process between the union and the firm's managers. Specifically, the manufacturing multi-output firm plays a role of leadership and pattern-setter for the other firms (and for the government), while both agricultural employment and unemployment have a residual role.

The institutional setup of the European countries is such that the role of unions is more relevant than elsewhere. The union is considered to be an agent who aggregates the preferences of those who participate in the labor market and who use their bargaining power to obtain a wage above the competitive level. The presence of unions provides the rationale for involuntary unemployment which has been a typical phenomenon of European economies in the 1980s. Bargaining is a sequential process: at first, unions and representatives of the manufacturing firm agree upon the wage rate; subsequently, the other firms and the government set their wages on the basis of differentials which depend upon the union bargaining power as well as upon general and sector-specific economic conditions (Dobson, 1994). On the other hand, there are no unions in the secondary labor market (i.e., agriculture) and wages are determined according to an arbitrage relative to the unemployment status.

To describe the process leading to the determination of equilibrium wages in the economy, we start from the manufacturing firm which, as said, is pattern-setter in the above process. This firm determines the level of employment on the basis of its labor demand function, taking as given the gross wage, which includes both the net wage paid to employees and the fiscal wedge (social security contributions paid by the employer plus income taxes paid by the worker). The net wage is the outcome of the bargaining process between the union and the firm. The fiscal wedge is set by the government. As for the determination of the net wage, the outcome of the bargaining process is assumed to be adequately described by the solution of a Nash bargaining optimization problem in which the equilibrium wage depends crucially upon two parameters: the union's relative bargaining power and the union's relative preference for employment. In the model, the first is endogenized as a function of the unemployment rate, unionization rate, and migration flows. Other variables influencing the equilibrium net wage are the replacement rate of manpower, the level of profits per worker, the wage elasticity of labor demand, and the fiscal wedge itself. This last variable is crucial as the size of its effect on the net wage provides a measure of the so-called wage resistance; i.e., how much unions increase the net wage in response to a reduction in the fiscal wedge (specifically, a reduction of payroll taxes through recycling of environmental tax revenue).

The wage equation has been estimated as a nonlinear Error Correction Mechanism in which the fiscal wedge affects wages both in the short run and in the long run (Brunello, 1996). In the long run, the estimated effect reflects, in all EU countries, the theoretical prediction that a 1% reduction of the wedge is completely offset by an increase in the net wage (the long-run elasticity is equal to −1). In the short run, the size of wage resistance is estimated as follows. A 1% reduction in the average payroll tax increases the net wage by 0.29% at the EU level. There is, however, a large variability across EU countries, because this effect goes from 0.48% in The Netherlands to 0.042% in Denmark. As far as the gross wage is concerned, a 1% reduction in the average payroll tax reduces the gross wage by 0.09% at the EU level (from 0.018% in Denmark to 0.14% in France). The cumulated effect is 0.22% at the EU level, and ranges from −0.04% in Denmark to −0.34% in France. These results suggest that cuts in payroll taxes may have a limited effect on employment because unions succeed in transforming part of the tax cut into higher net wages.

Once the equilibrium wage rate of the leading sector has been determined, the wage rates of the other sectors operating in the primary labor market (electricity and energy firms and government sector) are obtained simply by applying intersectoral wage differentials. The description of the functioning of the labor markets in the WARM model is then completed by noting that, given the wages, employment is determined by the labor demand equations (see the following section for the government's demand for labor). Given an exogenous temporal profile of the labor force, the rate of unemployment is determined residually.

The Government Sector

A noticeable feature of WARM is the endogenization of government behavior. The fundamental assumption here is that its strategies can be endogenized as functions of the economic policy goals, that is through a set of reaction functions.

The government undertakes four main activities. The first is the production of a public good (public spending) using inputs from labor, fixed capital, and a composite intermediate material good. This activity is formally represented by a variable Cobb–Douglas cost function which, via Shephard's Lemma, yields the demand for public employment and for materials. Both demand equations depend upon factor prices and are conditional upon a given amount of public good produced and of the stock of public capital.

The second economic activity is public investment. The stock of public capital depends upon past investment. However, unlike the behavior of the other agents, the optimal investment decision is not based on the government's variable cost function via the envelope condition. Instead, the assumption is made that public investment is one of the government's control variables used to smooth business-cycle fluctuations. It follows that public investment is endogenized through a reaction function which responds to changes in three factors: the cost of investment, the cost of funds available for financing investment expenditures, and the economic policy targets (captured by the rate of change in total employment).

The production and investment activities undertaken by the public agent are subject to the usual budget constraint whereby expenditures are financed by indirect and direct tax revenues and by issue of government bonds. The items of the budget currently endogenized in the model are the revenues from direct income taxation, the interest payments on the stock of public debt outstanding, the amount of pensions paid to households, and the social security contributions due from both households and firms. The other two activities of the government are therefore the regulation of the level of business activity through appropriate fiscal and expenditure policies, and the redistribution of income.

Looking now at the sources of public funds, indirect fiscal revenues are calculated by applying the effective tax rate and excise rate respectively to the value of output consumed and to the volume of goods and services consumed. Adding the revenues from each taxable basis, the overall fiscal revenue is obtained. The model accounts for 11 different VAT rates and 19 different excise rates. The revenues from direct taxation are computed on the basis of a single time-varying tax rate applied to total income.[15] The tax rate is endogenous and determined by a reaction function that depends upon the rate of inflation (to account for fiscal drag phenomena), the amount of public good supplied (public spending to be financed), and the Maastricht goals (debt to GDP ratio and budget deficit to GDP ratio, to account for the government's efforts in meeting the requirements for eligibility to the EMU).

Turning now to the uses of public funds, the nominal rate of interest on public debt is endogenized via a reaction function that depends upon the rate of inflation, the trade balance, and the growth rate of public debt (to capture the risk premium demanded by investors in high-debt countries). The amount of pensions paid by the government is related to the number of persons over 65 years relative to total population, to the average personal income, and to the distribution of income between wages and profits. Social contributions (as a share of total labor costs) depend upon the real wages, inflation rate and upon the average pension perceived by individuals.[16]

The Foreign Sector

The model adopts a production-theoretic view of international trade flows. In fact, households cannot make direct purchases from abroad and all flows of goods and services go through the production sector (especially the manufacturing firm) which performs repackaging, distribution, and similar activities.[17] Viewing imports as an input to the production process, this approach allows an integrated treatment of trade flows, in that exports are viewed as an output of the same production process (Kohli, 1991).

   The model is characterized by a precise specialization in the import–export activity of firms. The agricultural firm can sell its output to both domestic and foreign markets. The manufacturing firm is the only agent trading manufactured goods and services internationally. Imported electricity is only handled by the corresponding firm and this is also the case of other energy sources for the corresponding firm.

   Import decisions take place through a two-stage allocation process. The procedure is the same as the one previously described for generic inputs: on the basis of relative prices the total import demand for each good and service is determined. Next, the aggregate price of a good being bought by the domestic firm from abroad is viewed as a unit cost function. Shephard's Lemma then yields the demand function of that good by the importing country from each selling country. Analytical tractability for a problem that involves many goods and many countries at the same time suggested the use of Cobb–Douglas unit cost functions.

   On the export side, the approach used describing the behavior of producers is followed here as well. The price of a good produced by country i and sold to country j is a constant fraction of the production price of the same good. The simple representation of trade flows just outlined is sufficient for model the structure of bilateral exchanges of goods, services, and energy sources among the EU12 countries as well as the "rest of the world."

Emissions

The role of the environmental module in WARM serves the purpose of assessing the impact of the economic activities on the environment. This task is performed by measuring the amount of harmful emissions.[18] A series of technical coefficients provide the amount of emissions generated by the various forms of economic activity specified by the model. The attention is centered upon the following pollutants: sulfur oxide (SO_x), nitrogen oxide (NO_x), particulate matter, carbon monoxide (CO), and volatile organic components. These pollutants are to a large extent related to the use of fossil fuels in the various production and consumption activities. Moreover, fossil fuels are the most relevant anthropogenic source of CO_2 emissions, which are largely responsible for the greenhouse phenomenon. The model disaggregation across economic activities (i.e., economic agents) and across energy sources allows a precise assessment of individual contributions to overall emissions.

3. Environmental Fiscal Reforms under Alternative Institutional Settings

The model just described was used to carry out some simulation experiments regarding the so-called "employment double dividend." This hypothesis suggests that an appropriately designed fiscal reform, in which emission charges are used to subsidize employers' social security contributions, may reach (at least) two relevant policy goals: a better quality of the environment and, at the same time, an increase in employment levels. This issue, which has been studied elsewhere (Carraro et al., 1996) has recently heated the debate on the effects of carbon taxation, especially in Europe. Here we take it up again and analyze it under alternative possibilities concerning the institutional level at which the policy decision is made. The goal of the following experiments is therefore twofold: (1) to verify whether the "employment double dividend" emerges across different institutional settings, and (2) to compare the results thus obtained in order to quantify gains and losses (in terms of GDP growth, employment, and emissions) of those alternative institutional scenarios.

As stated in the Introduction, the process of European integration modifies the role and effectiveness of monetary and fiscal policies and of environmental policies as well. In particular, the effects of an environmental fiscal reform designed to curb emissions and to boost employment may differ according to the degree of harmonization with which it is introduced. A high level of cooperation among EU countries may increase the benefits of the reform in terms of employment and reduce its costs in terms of emission abatement. On the one hand, free-riding and intra-Europe leakage are likely to be lower; on the other hand, the fiscal revenue can be redistributed across EU countries according to its marginal effectiveness in reducing unemployment.

To analyze these issues we ran three simulation experiments. In the experiments, energy taxation increases in order to control carbon emissions whereas the tax revenue is recycled through a generalized reduction in payroll taxes in order to stimulate employment. In the baseline scenario, the EU economy as a whole exhibits a steady-state growth at an annual rate of 2.4%, with a stable inflation rate of 1.8%. It is assumed that the fiscal reform is introduced in 1995. Simulations are carried out up to the year 2010. The environmental fiscal policy just described can be decided at three different institutional levels which are now considered in turn.

Harmonized Tax Rates with Domestic Revenue Recycling

In this first experiment, it was assumed that all EU countries agree to introduce a harmonized environmental policy characterized by identical tax rates across countries, thus equalizing marginal abatement costs. In practice, the carbon tax is introduced according to the European Union proposal (*Official Journal of the EC*, 196/1, 1992). In particular, the proposal specifies the tax as a mixed charge on emissions and on the energy content of different fuels and of electricity. We introduced the tax at the highest level proposed by the European Commission (19 ecu per ton of CO_2, which corresponds to about $10 per barrel). The tax is levied on manufacturing and final energy consumption. The revenue of the tax is used to reduce payroll taxes, which are gross labor costs which enter as an argument in labor demand equations. Notice that in this case the tax rates are harmonized (e.g., as the outcome of an environmental cooperative agreement), whereas the revenue recycling takes place in every country under the assumption that each domestic budget deficit (surplus) remains unchanged.

The impact of the harmonized fiscal reform on gross nominal wages, employment, GDP growth, and CO_2 emissions is shown in Table 1 for six EU countries (France,

Table 1. *Effects of a Harmonized Tax*[a]

	France		Germany		Italy	
	Short run	*Long run*	*Short run*	*Long run*	*Short run*	*Long run*
Tax rate (%)	19	19	19	19	19	19
Gross nominal wages	−2.15	1.83	−3.06	3.91	−2.32	2.44
Employment	0.76	−0.22	0.50	−0.08	0.39	0.04
Real GDP growth	−0.55	0.32	−1.18	0.21	−0.61	0.03
CO_2 emissions	−4.17	5.71	−4.25	−2.21	−2.28	5.72
	The Netherlands		Spain		United Kingdom	
	Short run	*Long run*	*Short run*	*Long run*	*Short run*	*Long run*
Tax rate (%)	19	19	19	19	19	19
Gross nominal wages	−4.73	0.45	−2.84	5.42	−2.01	5.82
Employment	3.74	0.27	1.21	0.44	0.25	1.78
Real GDP growth	−0.38	0.11	−0.65	0.39	−0.72	0.36
CO_2 emissions	−6.49	−4.23	−2.53	2.31	−2.35	2.01

[a] Percentage differences with respect to the baseline. Short run is after two years; long run is in year 2010.

Germany, Italy, The Netherlands, Spain, and the United Kingdom).[19] The qualitative impact of the fiscal reform is basically the same in all countries. In the short run, wages are reduced and employment increases. However, in the medium run (after about 10 years) wages tend to revert back to their baseline level and to increase afterwards. As a consequence, the initial increase of employment disappears in the long run.

This evidence can be rationalized in terms of the modeling of the labor market: wage bargaining implies that only in the short run are unions unable to offset the payroll tax reduction through an increase of net wages. In the long run, the fiscal change is completely absorbed by the change of net wages. Hence gross wages, other things being equal, go back to the initial level. They further increase with respect to the baseline scenario because of two feedback mechanisms: (1) the unemployment reduction achieved in the short run increases unions' bargaining power, thus inducing, through the dynamics of the bargaining equation, a counter-effect in the long-run; and (2) the energy-saving R&D activity induced by the energy tax increases the growth rate of technical progress, thus increasing factor productivity and, as a consequence, firms' profits. Since the wage bargaining equation is in the WARM model formulated in terms of gross wages as a proportion of profits, in the long run gross wages are larger than in the baseline.

A noticeable exception is the United Kingdom, where there is a positive effect on employment even in the long run. This can be explained by the higher bargaining power of unions in that country and by their relatively higher preference for employment. Unions are therefore able to achieve both a small reduction in gross wages and an increase in employment levels.

The simulated fiscal reform can therefore achieve its main goal (employment increase) only in the short run. After that, there is no further employment gain that can be reaped by the fiscal maneuver.[20]

As far as the effects of the fiscal reform on output growth and CO_2 emissions are concerned, Table 1 shows that the long-run effect on output growth is slightly above zero. In the short run, an output effect prevails; that is, the tax change increases costs and reduces output. However, in the long run, the initial positive shock on employment induces an increase in aggregate demand (through households' higher income and consumption) which in turn fosters GDP growth. However, this growth obviously produces negative effects on emissions.

Therefore, as expected, the fiscal reform reduces emissions in the short run even though the beneficial effect on emissions tends to disappear in the long run. The transitory emission reduction can be explained by looking at the effect of the fiscal reform both on the level and the composition of aggregate demand. As previously said, unions tend to capture part of the rent given to firms by the reduction of payroll taxes. Hence, gross wages decrease less than expected, whereas net wages increase. In the long run, all the reduction in payroll taxes is transformed into a net wage increase. The increase in net wages and in employment induces an increase in households' income, and then in the consumption of all goods, including energy.[21] As technical progress mainly affects production technologies, emissions per unit of output decrease, but emissions per unit of consumption do not. Therefore, the increase in energy consumption leads to higher emission levels. In some countries, this revenue effect can be larger than the emission reduction induced by higher energy prices (the substitution effect).

In summary, the empirical analysis seems to suggest that no "employment double dividend" can emerge in the long run. By contrast, the fiscal reform may lose the environmental dividend. There is therefore evidence in favor of the usual long-run environment/employment tradeoff. However, in the short run the situation may be different, as employment increases and emissions decrease, thereby indicating the possibility of a short-run "employment double dividend."

Decentralized Tax Rates: An Assessment of the Subsidiarity Principle

Let us now consider the case where the reform is introduced in all 12 EU member states, but where, according to the subsidiarity principle, both the tax rate and the revenue recycling are set independently by each national government. In this case, each government's optimal fiscal reform would result from the maximization of his own welfare function with respect to the emission tax rate and to the level of social security contributions. The definition of a welfare function consistent with the theoretical framework described above is a quite complex task, in particular if separability between environmental and economic variables cannot be considered a realistic assumption. Moreover, the international economic and environmental spillovers would require the optimal fiscal reform in a given country to account for the decisions taken in the other countries; i.e., a policy game should be solved explicitly. This task is too complex with a model of about 10,000 equations such as WARM. Nonetheless, we believe that a reasonable approximation of the optimal policy in the case in which the subsidiarity principle is applied can be attained by the following scenario.

Countries with a low environmental damage and/or high energy tax rates sets a lower CO_2 emission tax rate, and vice versa. In particular, we take the emissions/output ratio as a rough proxy of the technological characteristics of each country which summarizes information that should come from the environmental damage function. Even if this is just an approximation, the idea is that countries with a lower emissions/output ratio perceive a lower environmental damage which leads them to set a lower

Table 2. Effects of a Decentralized Tax[a]

	France		Germany		Italy	
	Short run	Long run	Short run	Long run	Short run	Long run
Tax rate (%)	17	19	21	20	15	12
Gross nominal wages	0.20	0.13	−0.35	0.21	0.36	−0.05
Employment	−0.05	−0.06	0.04	−0.02	−0.07	−0.04
Real GDP growth	0.04	0.04	0.04	−0.07	0.06	0.02
CO_2 emissions	0.30	−0.40	−0.40	0.00	0.40	−1.40

	The Netherlands		Spain		United Kingdom	
	Short run	Long run	Short run	Long run	Short run	Long run
Tax rate (%)	20	17	21	16	20	25
Gross nominal wages	−0.34	0.20	−0.16	0.32	−0.04	1.12
Employment	0.23	−0.10	0.07	0.03	−0.15	0.35
Real GDP growth	−0.02	0.02	−0.05	0.03	−0.05	0.36
CO_2 emissions	−0.40	0.20	−0.18	0.26	−0.20	0.00

[a] Percentage differences with respect to harmonized case. Short run is after two years; long run is in year 2010.

tax rate. Moreover, the existing level of energy taxation is used as a measure of the distortionary features of the fiscal system (on this issue, see Bovenberg (1997) and Carraro and Soubeyran (1996)).

The revenue recycling takes place at the national level; that is, each government uses the tax revenue to reduce employers' payments for social security contributions under the constraint that the domestic budget surplus (deficit) remains unchanged.

The results of this experiment are shown in Table 2. It can be seen that Germany and the UK introduce a tax rate that is higher than the European average, whereas Italy and France impose a lower one (the European average is 19 ecu per ton of CO_2). These two countries suffer from a reduction in employment levels, with respect to the harmonized tax rate case analyzed above, both in the short run and in the long run. In The Netherlands and Spain, where the tax rate is initially relatively higher but decreases in the longer run, employment goes up in the short run but gets then reduced. There thus emerges a negative correlation between tax rate and employment, with respect to which the UK again represents an exception. In general, tax rates and emissions also appear to be negatively correlated (again recall that all results are shown with respect to the harmonized case of Table 1). These correlation pattern between tax rate, employment, and emissions are in line with our *a priori* expectations.

From the analysis it can thus be concluded that decentralizing the decision about tax rates may lead some countries to set too low rates, thus reducing the effect of the reform on employment levels. In fact, the benefits of the environmental fiscal reform come mainly from revenue recycling; i.e., from lower gross wage levels. However, if tax rates are too low, the revenue to be recycled is not large enough to induce significant changes of gross wages and then of employment.

An Environmental Fiscal Reform with Federal Revenue Recycling

Another interesting institutional setting is the one in which European integration is
fully achieved. In this case, we assume that a central government (e.g., the European
Commission) defines the optimal tax rate and the optimal social security contribution
policy in each country. In the absence of a European welfare function, we approximate
this case with the following scenario. Let us assume that the tax rate is homogeneous
across countries (to equalize marginal abatement costs), whereas the tax revenue is
recycled by giving proportionally more resources to those countries with higher unemployment levels (the federal budget surplus or deficit remains unchanged). Therefore,
the reduction of payroll taxes is larger in these countries. The idea is that the European
Commission wants to redistribute the fiscal revenue from the emission tax in a way
which minimizes the social costs of unemployment (which are assumed to be larger
where unemployment is larger).

    From Table 3 it can be seen that resources (tax revenues) go in the short run from
Germany and The Netherlands to the other countries, especially Italy and Spain. These
two countries can therefore benefit from significant increases in both GDP growth and
employment levels.[22] The opposite happens in those countries which finance such
resource transfer, Germany in particular. As time goes by, however, economic integration sets forth a re-equilibrating process, so that the growth and employment gains
realized by Italy and Spain get reabsorbed, and the same happens for the growth and
employment losses of the other countries. Actually, Spain receives in the short run the
largest transfer and scores such a big increase in both employment and GDP growth as
to become in the longer run a net contributor of revenues to the federal budget, along
with the UK. Of this fact, France takes special advantage.

Table 3. The Effects of a Federal Tax[a]

	France		Germany		Italy	
	Short run	*Long run*	*Short run*	*Long run*	*Short run*	*Long run*
Net revenue	−500	6200	−5000	−800	5800	8700
Gross nominal wages	0.02	−0.52	−0.12	0.20	−0.42	−0.48
Employment	−0.02	0.14	−0.19	0.06	0.24	0.08
Real GDP growth	−0.06	0.00	−0.48	0.30	0.24	0.00
CO_2 emissions	−0.02	0.04	−0.12	0.04	0.04	−0.03

	The Netherlands		Spain		United Kingdom	
	Short run	*Long run*	*Short run*	*Long run*	*Short run*	*Long run*
Net revenue	−4100	−4000	7500	−3400	−500	−8600
Gross nominal wages	0.72	−0.30	−2.00	−0.30	−0.02	−0.60
Employment	−0.56	0.20	1.40	−0.30	−0.28	−0.56
Real GDP growth	−0.08	−0.02	0.58	0.02	−0.05	0.14
CO_2 emissions	−0.02	−0.05	0.28	−0.10	0.02	−0.58

[a] Percentage differences with respect to harmonized case. Short run is after two years; long run is in year
2010. Net revenue is expressed in millions of ecu.

Table 4. Effects of the Three Policy Scenarios on Aggregate European Variables[a]

	Harmonized		Decentralized		Federal	
	Short run	Long run	Short run	Long run	Short run	Long run
Gross nominal wages	−2.25	0.87	−1.98	1.13	−2.29	0.94
Employment	0.78	0.09	0.53	0.07	1.45	0.24
Real GDP growth	−0.45	0.14	−0.35	0.13	−0.39	0.12
CO_2 emissions	−3.87	−0.18	−3.14	0.06	−3.81	−0.08

[a] Percentage differences with respect to the baseline. Short run is after two years; long run is in year 2010.

On the whole, there clearly emerges a positive correlation between net revenue transfer to a country and gains in terms of GDP growth and employment and increases in emissions. The advantages of a federal environmental fiscal reform appear to concern employment above all. On the one hand, the reform redistributes employment in the short run; on the other hand it helps increasing employment in the long run. By contrast, no significant change in terms of emissions or GDP growth seems to appear. This is true for four out of the six countries examined.

A further assessment of the effects of a centralized environmental fiscal reform can be provided by comparing the value of aggregate European employment and emissions in this case with their value in the first two institutional settings analyzed in this paper (the case where only emission tax rates are harmonized and the case in which the subsidiarity principle is applied).

From Table 4, in which differences are shown with respect to the baseline scenario, it can be noticed that the largest impact on employment is achieved when a federal redistribution policy is implemented. Even if the effects on employment slow down in the long run (because of the functioning of the labor market described above), a cooperative setting of both emission and employment policies enables the European Commission to achieve a significant impact on employment levels without excessively penalizing the environment in the long run (significant short-run reduction of CO_2 emissions is achieved).

By contrast, the figures in the case in which the subsidiarity principle is applied show that employment and environmental benefits are smaller than in the other cases. The small differences with respect to the harmonized emission rate case can be explained by the similarity of economic and fiscal structures in Europe. As mentioned above, only France (because of its lower levels of CO_2 emissions), Italy (because of its high tax rates on energy), and the UK (because of the high coal consumption) introduce a fiscal reform with significant differences from the average European one.

Conclusions

This paper has used a new econometric general equilibrium model for 12 countries of the European Union to analyze the effects of a fiscal reform designed to achieve, at the same time, a better protection of the environment (e.g., a stabilization of CO_2 emissions) and a lower tax burden on labor so as to increase employment. The empirical analysis was carried out under three alternative institutional settings: a fiscal reform where emission tax rates are harmonized but the recycling policy is set at the national level; a decentralized tax and revenue recycling policy which mimics the application of the subsidiarity principle; and a federal fiscal reform where a central authority coordinates both the tax rates and the employment policy. Although they are very preliminary, the results provide some useful indications which can be summarized as follows.

First, in general the effects of an environmental fiscal reform designed to reap an environment–employment "double dividend" depend mainly upon the functioning of the labor market. Given the centralized wage bargaining characterizing many EU labor markets, it is unclear whether wage subsidies (i.e., reductions in payroll taxes) are the appropriate instruments to stimulate employment. Our results show that an increase of emission charges and a reduction of payroll taxes reduce pollution and increase employment mainly in the short run. In the long run, the usual tradeoff between environment and employment seem to emerge.

Second, modifying the institutional setting does not change the qualitative features which define the effects of the fiscal reform and only slightly modify its quantitative effects. However, if the main goal is employment relief, rather than emission reduction, the results suggest that a federal policy in which both tax rates and wage subsidies are harmonized is the best institutional setting. Tax rates have to be harmonized because decentralization would lead to low rates which provide less resources to be recycled to boost employment. Wage subsidies have to be coordinated because revenue recycling produces the largest increase of EU employment when the tax revenue is used chiefly in those countries with a high unemployment record.

Third, the small effects on employment of the environmental fiscal reform suggest that limiting the reform to emission charges and wage subsidies is probably inefficient. The idea of reforming the fiscal system by correcting externalities and by lowering distortionary taxes should be considered in more general terms, by proposing an overall change of the tax system. A comprehensive, well-designed reform is likely to provide larger gains in all countries.

Finally, it seems that the environmental fiscal reform analyzed in this paper can provide significant benefits only in the short run. How to achieve permanent gains in employment and GDP growth is an open question. An answer may be provided by policies which stimulate both human and physical capital accumulation, such as R&D and innovation policies.

It is important to conclude these comments with a word of caution in interpreting the results. Gains and benefits of the environmental fiscal reform should be evaluated in terms of welfare changes, even though policymakers in Europe are actually interested in the dynamics of some specific variables such as unemployment. The definition of a welfare function would also enable us to compute the optimal noncooperative and cooperative fiscal reforms. In addition, political economy considerations should not be neglected. Nonetheless, we believe that the scenarios that we have designed to approximate the three institutional settings to be analyzed captures their main features. As a consequence, our results provide sufficiently reliable, albeit preliminary, information, on the effects of an environmental fiscal reform in an integrated Europe. More research, both theoretical and empirical, is clearly necessary to better understand the qualitative and quantitative aspects of the policy issues discussed in this paper.

Appendix: The Endogenous Technical Progress in WARM

One of the crucial features of the WARM model is an explicit attempt at describing the mechanism through which economic variables affect technical change. The basic idea is that the dynamics of technical change cannot be observed in contrast with traditional approaches in which technical change is proxied either by a time trend or by the computation of factor productivity indexes.[23] We adopt a latent variable structural equation model which uses data on total expenditures in R&D from public and private sources, on imports of patents, and on business cycle indicators as cause variables for the latent technological variable. The latent variable approach extracts information from indicators and cause variables while being able to avoid using them as exact representations of technological change.

To be specific, we assume that the capital stock is made up of two components: an energy-saving (or energy-efficient), environment-friendly stock and an energy-consuming, polluting one. More precisely, two technologies are assumed to be available to produce the same output: the traditional one, and the best-available environmental technology. The physical capital stock is therefore formed by assets which embody either the first or the second type of technology. Each year a new vintage of capital becomes operational. In this way new capital is added to the two components. The composition of this new capital depends on a number of economic variables which affect the firms' decision of installing environment-friendly capital (e.g., the energy prices).[24] The quality of the environmental technology is affected by the amount of R&D spending carried out by firms.

Let k_t be the total capital stock and k_e and k_p the environment-friendly and polluting stocks respectively. By definition, $k_t = k_e + k_p$, which implies

$$g_t = g_p + (g_e - g_p)(k_e/k_t),$$ (A1)

where g_t, g_p, and g_e are the growth rates of the overall, polluting, and environment-friendly capital stocks respectively. Suppose that

$$g_e - g_p = f(X)(k_t/k_e) + \varepsilon,$$ (A2)

where $f(X)$ is the capital growth rate in the long run, when all technological possibilities to reduce energy consumption have been actually implemented; i.e., when $k_t = k_e$ and $g_p = 0$; X is a set of explanatory variables, and ε is a stochastic error.[25] The implicit assumption here is that when the stock of polluting capital is high the rate of growth of the environment-friendly capital is larger than the rate of growth of the polluting capital; the difference decreases as k_e approaches k_t. Finally, the following equation defines the dynamics of the polluting component of the capital stock:

$$g_p = h(W, v),$$ (A3)

where W is a set of explanatory variables and v is a stochastic error term. In particular, the explanatory variables include R&D spending, output demand, factor prices, and the number of imported patents. Everything else equal, it is likely that more R&D spending increases the technological possibilities of the economic system, thereby inducing investment in environment-friendly capital which replaces investment in polluting capital. Similarly, higher energy prices may induce firms to reduce investment in energy-consuming technologies.[26]

The amount of R&D carried out by firms is an endogenous variable of the model. We relate it to total output demand (assuming a unitary elasticity in the long run), relative factor prices, and policy variables. These include environmental taxation (via energy prices) and innovation subsidies (via publicly funded R&D expenditures).

The three equations (A1)–(A3) and the set of equations which endogenize R&D expenditures, factor prices, and output demand define the structure of the latent variable model. As g_p and g_e are not observable, they must be estimated by filtering the information contained in the observable variables. To do that, let us rewrite expressions (A1)–(A3) in a (linear) state–space form as follows:

$$g_e = Hs + \varepsilon, \tag{A4}$$

$$s = Fs(-1) + v, \tag{A5}$$

where s is the state space vector which contains the unobservable variable g_p and the parameter vectors β and δ associated with the variable vectors X and W, respectively. More precisely:

$$H = \begin{bmatrix} 1 & X & 0 \end{bmatrix}, \quad s = \begin{bmatrix} g_p \\ \beta \\ \delta \end{bmatrix}, \quad F = \begin{bmatrix} m & 0 & W \\ 0 & 1 & 0 \\ 0 & 0 & 1 \end{bmatrix}.$$

The matrix H is called the output matrix, while F is the transition matrix of the state–space form of the model; the latter includes the parameter m which captures the adjustment speed of the components of the capital stock.[27] The state–space form (A4)–(A5) has been estimated using the square-root Kalman and information filters described by Carraro (1988).[28] The empirical results are briefly summarized in Table 5.

In the first column we report the coefficients capturing the speed of adjustment m of the composition of the capital stock to the desired value (the only unknown parameter in the F matrix); the second column reports the autonomous change of the growth rate of the polluting capital stock g_p (the constant term in the vector X); the third and fourth columns refer to the change in g_p induced by domestic R&D expenditures and by imported patents; while the last column contains the impact on the same growth rate of output growth.

Notice that, as expected, domestic R&D and imported patents reduce the growth rate of the polluting capital stock, which is therefore replaced by environment-friendly capital. By contrast, when output grows both types of capital stock grow. Finally, the speed of adjustment is quite high in all countries, thereby showing little sluggishness in

Table 5. *Estimation Results of the Kalman Filter Technique*

Country	Speed of adjustment	Autonomous change	R&D	Patents	Output
Belgium	0.807	−0.0037	−0.0226	−0.0214	0.209
Denmark	0.924	−0.0024	−0.0226	−0.0214	0.210
France	0.726	−0.0030	−0.0026	−0.0216	0.209
Germany	0.857	−0.0017	−0.0026	−0.0217	0.209
Greece	0.817	−0.0082	−0.0027	−0.0215	0.209
Ireland	0.924	−0.0082	−0.0027	−0.0218	0.209
Italy	0.842	−0.0034	−0.0026	−0.0215	0.209
Luxembourg	0.720	−0.0012	−0.0025	−0.0222	0.209
Netherlands	0.849	−0.0030	−0.0026	−0.0214	0.209
Portugal	0.887	−0.0017	−0.0029	−0.0219	0.209
Spain	0.885	−0.0044	−0.0024	−0.0216	0.210
UK	0.869	−0.0012	−0.0027	−0.0217	0.209
EU12	0.768	−0.0026	−0.0026	−0.0217	0.209

environmental innovation. Relevant differences in the coefficients across EU countries can be found only for the autonomous change of the growth rate of the polluting capital stock, which decreases more rapidly in Spain, Greece, and Ireland; i.e., in less developed countries. As more developed countries have already implemented a large number of best available technologies, the substitution between the two components of the capital stock takes place more slowly in such countries (the best examples are Germany and the UK).

On the basis of the above results, the dynamics of the two time series k_e and k_p has been reconstructed.[29] Then an indicator of technical change, here interpreted as an indicator of the environmental quality of the capital stock, is provided by the ratio $T = k_e/k_p$. The average growth rate of this indicator is fairly low in the EU developed countries (about 2% in Germany, France, and Italy, slightly lower in the UK), whereas it is much higher in the less developed countries (from 9% in Ireland to 30% in Greece). In all countries, the growth rate of the technical progress indicator becomes lower as the country grows (because the model objective is to capture the implementation of best-available technologies in the short-run and medium-run).

The dynamics of $T = k_e/k_p$ generally affects the decision rules of all the agents in the model: not only does it affect the firms' input and output decisions, as previously shown, but it also influences households' choices, especially those pertaining to the consumption of energy sources. Moreover, the dynamics of technical change concern the total private capital stock, and not only the component used as a production input by firms.[30] Note that an important implication of our treatment of environmental technical change is that it is a diffuse phenomenon. In fact, the technological indicator is an argument of the behavioral equations concerning the demand for inputs and the supply of outputs (via marginal costs). In particular, this is the case for the manufacturing firm. Thus, over time an increasing amount of environment-friendly capital is used in production which translates into an improvement of the production process and in turn of the quality of goods and services supplied. To the extent that these goods and services are purchased by the other firms and by households, the process of endogenous technical change affects all sectors of the economy (and the rest of the world as well).

References

Boone, Lawrence, Stephen Hall, and David Kemball-Cook, "Endogenous Technical Progress in Fossil Fuel Demand: The Case of France," Centre for Economic Studies discussion paper 21-93, 1992.

Bovenberg, Lans, "Environmental Policy, Distortionary Labour Taxation and Employment: Pollution Taxes and the Double Dividend," in Carlo Carraro and Domenico Siniscalco (eds.), *New Directions in the Economic Theory of the Environment*, Cambridge: Cambridge University Press, 1997.

Bovenberg, Lans and Larry Goulder, "Integrating Environmental and Distortionary Taxes: General Equilibrium Analysis," paper presented at the Conference on Market Approaches to Environmental Protection, Stanford University, 3–4 December 1993.

Bovenberg, Lans and Rick Van der Ploeg, "Does a Tougher Environmental Policy Raise Unemployment? Optimal Taxation, Public Goods and Environmental Policy with Rationing of Labour Supply," CEPR discussion paper 869, 1993.

———, "Environmental Policy, Public Finance and the Labour Market in a Second-Best World," *Journal of Public Economics* 55 (1994a):349–90.

———, "Green Policies and Public Finance in a Small Open Economy," *Scandinavian Journal of Economics* 96 (1994b):343–63.

Bovenberg, Lans and Rick Van der Ploeg, "Optimal Taxation, Public Goods and Environmental Policy with Involuntary Unemployment," *Journal of Public Economics* 62 (1996):59–83.

Brunello, Giorgio, "Labour Market Institutions and the Double Dividend Hypothesis," in C. Carraro and D. Siniscalco (eds.), *Environmental Fiscal Reform and Unemployment*, Dordrecht: Kluwer Academic, 1996.

Buiter, Willem. H. and Richard. C. Marston, *International Economic Policy Coordination*, Cambridge: Cambridge University Press, 1985.

Capros, Pantelis, et al., "Double Dividend Analysis: First Results of a General Equilibrium Model Linking the EU-12 Countries," in C. Carraro and D. Siniscalco (eds.), *Environmental Fiscal Reform and Unemployment*, Dordrecht: Kluwer Academic, 1996.

Carraro, Carlo and Marzio Galeotti, "WARM: A European Model for Energy and Environmental Analysis," *Environmental Modelling and Assessment* 2 (1996):171–189.

Carraro, Carlo and Francesco Giavazzi, "Game Theory and Interdependence of Economic Policies," special issue of *Ricerche Economiche* 3–4 (1987):291–463.

Carraro, Carlo and Domenico Siniscalco, *Environmental Fiscal Reform and Unemployment*, Dordrecht: Kluwer Academic, 1996.

Carraro, Carlo and Antoine Soubeyran, "Environmental Taxation and Employment in a Multi-Sector General Equilibrium Model," in Carlo Carraro and Domenico Siniscalco (eds.), *Environmental Fiscal Reform and Unemployment*, Dordrecht: Kluwer Academic, 1996.

Carraro, Carlo, Marzio Galeotti, and Massimo Gallo, "Environmental Taxation and Unemployment: Some Evidence on the Double Dividend Hypothesis in Europe," *Journal of Public Economics* 62 (1996):141–81.

Conrad, Klaus, "Coordination vs Non-coordination of Environmental Policies in Europe," report prepared for the EU Project on Climate Change Strategies within Competitive Energy Markets, University of Mannheim, 1996.

Currie David and Paul Levine, "Macroeconomic Policy Design in and International World," in Willem H. Buiter and Richard C. Marston (eds.), *International Economic Policy Coordination*, Cambridge: Cambridge University Press, 1985.

Currie, David, Paul Levine, and Nic Vidalis, "International Cooperation and Reputation in an Empirical Two-Bloc Model," CEPR discussion paper 198, 1987.

Denis, Cecile and Gert J. Koopman, "Differential Treatment of Sectors and Energy Products in the Design of a CO_2 Energy Tax: Consequences for Employment, Economic Welfare and CO_2 Emissions," presented at the FEEM Conference on Environmental Taxation, Revenue Recycling and Unemployment, Milan, 16–17 December 1994.

Dobson, Alan, "Multifirm Unions and the Incentive to Adopt Pattern Bargaining in Oligopoly," *European Economic Review* 38 (1994):87–100.

Drèze, Jacques H. and Malinvaud, Edmond, "Growth and Employment: The Scope of a European Initiative," manuscript, CORE, Louvain, 1993.

Galeotti, Marzio, "The Intertemporal Dimension of Neoclassical Production Theory: A Survey," *Journal of Economic Surveys* 10 (1996):1–40.

Gao, Xen M., "Measuring Technological Change Using a Latent Variable Approach," *European Review of Agricultural Economics* 21 (1994):113–29.

Grubb, Michael, et al., "The Costs of Limiting Fossil-Fuel CO_2 Emissions," *Annual Review of Energy and Environment* 18 (1993):397–478.

Holmlund, Berndt and Jakob Zetterberg, "Insider Effects in Wage Determination," *European Economic Review* 35 (1991):1009–34.

Hourcade, Jean Charles, "Modelling Long-Run Scenarios: Methodology Lessons from a Prospective Study on a Low CO_2 Intensive Country," *Energy Policy* 21 (1993).

Hughes Hallett, Andrew and James Andrew, "How Much Could the International Coordination of Economic Policies Achieve? An Example from US–EEC Policy Making," CEPR discussion paper, 1985.

Hughes Hallett, Andrew and Yi Ma, "Changing Partners: The Importance of Coordinating Fiscal Monetary Policies within a Monetary Union," *Manchester School* (1995, forthcoming).

Hughes Hallett, Andrew and Peter McAdam, "Fiscal Deficit Reductions in Line with the Maastricht Criteria for Monetary Union: An Empirical Analysis," CEPR discussion paper 1351, 1996.

Jorgenson, Dale W. and Peter J. Wilcoxen, "Intertemporal General Equilibrium Modelling of US Environmental Regulation," *Journal of Policy Modelling* 12 (1990):715–44.

Kohli, Uhli, *Technology, Duality, and Foreign Trade*, Hemel Hempstead: Harvester Wheatsheaf, 1991.

Layard, Richard, Richard Jackman, and Stephen Nickell, *Unemployment*, Oxford: Blackwell, 1991.

Mélitz, Jacques and Axel Weber, "The Costs/Benefits of a Common Monetary Policy in France and Germany and Possible Lessons for Monetary Union," CEPR discussion paper 1374, 1996.

Morrison, Christine J., "Quasi-Fixed Inputs in US and Japanese Manufacturing: A Generalised Leontief Restricted Cost Function Approach," *Review of Economics and Statistics* 70 (1988):275–87.

Oates, Wallace, "Pollution Charges as a Source of Public Revenues," Resources for the Future, discussion paper QE92-05, Washington, DC, 1991.

Oudiz, Gilles and Jeffrey Sachs, "Macroeconomic Policy Coordination Among the Industrial Economies," *Brooking Papers on Economic Activity* 1 (1984):1–64.

——, "International Policy Coordination in Dynamic Macroeconomic Models," in Willem H. Buiter and Richard C. Marston (eds.), *International Economic Policy Coordination*, Cambridge: Cambridge University Press, 1985.

Pearce, David W., "The Role of Carbon Taxes in Adjusting to Global Warming," *Economic Journal* 101 (1991):938–48.

Slade, Margaret E., "Modelling Stochastic and Cyclical Components of Technical Change: An Application of the Kalman Filter," *Journal of Econometrics* 41 (1989):363–83.

Veugelers, Reinhilde and Hylke Vandenbussche, "European Anti-Dumping Policy and the Profitability of National and International Collusion," CEPR discussion paper 1469, 1996.

Notes

1. Examples of the early literature on the gains from the international coordination of economic policies are Hughes-Hallett and Andrew (1985), Currie et al. (1987), Oudiz and Sachs (1984, 1985), and the volume edited by Buiter and Marston (1985).

2. Conrad (1996) is one of the very few examples we are aware of.

3. Among the papers which have addressed this issue are Pearce (1991), Oates (1991), Bovenberg and Van der Ploeg (1993, 1994a, 1994b, 1996), and the studies collected in the volume edited by Carraro and Siniscalco (1996).

4. See for instance the papers contained in part II of Carraro and Siniscalco (1996).

5. In addition, the existing applied general equilibrium or econometric models for environmental policy analysis do not appear to satisfactorily address some fundamental aspects of environment–economy linkages (Grubb et al., 1993; Hourcade, 1993). More importantly, in order to properly assess the validity of the "employment double dividend" hypothesis, it is necessary to capture the crucial features of the European labor market, which is highly centralized and in which wages are usually the outcome of a bargaining process between unions and firms (Layard et al., 1991; Holmlund and Zetterberg, 1991). However, most empirical models, being designed to represent above all the energy sector, do not incorporate a satisfactory description of the labor market, which is often assumed to be perfectly competitive.

6. This model has been used to get a first assessment of the effects of an environmental fiscal reform of the type proposed here in Carraro et al. (1996).

7. The WARM model is the outcome of a large research project funded by the European Commission, DGXII-E1, and by the Fondazione ENI Enrico Mattei of Milan. The research work has been carried out at GRETA, Venice. Owing to data availability, the model is currently limited to the description of real flows, and only contains a broad disaggregation of production activities.

8. This firm carries out economic activities which go beyond manufacturing *strictu sensu*, as it also produces services. We term it "manufacturing" for the sake of brevity.

9. While it should be noted that the perfect competition hypothesis may be sustained in the case of (very) long-run analyses, entry barriers which do not vanish in the medium term can be considered as the rationale for imperfect competition in this model. Moreover, from the European experience it is apparent that labor markets are far from competitive. Finally, in WARM some activities are noncompetitive by construction: for instance, this is the case of the electricity and energy firms which are the sole domestic suppliers of their output.

10. Notice that outputs are expressed at factor cost, so that the corresponding prices are net of (net) indirect taxes. These tax rates are introduced when specifying the demand price for each product.

11. This is the first-order condition for the problem of minimizing total short-run costs, given by the sum of minimized variable costs and expenditures in fixed (i.e., capital) inputs (see, e.g., Galeotti, 1996).

12. Although the WARM model includes an endogenous labor supply decision, the simulations reported below assumed an exogenous labor supply (see Carraro and Galeotti (1996) for more details).

13. The idea of treating technical change as an unobserved or latent variable is shared by the partial equilibrium approaches of Slade (1989) and Gao (1994), and by the general equilibrium model of Boone et al. (1992).

14. From the brief description provided in the text it should clearly emerge that the WARM model does endogenize technical change but does *not* provide a structural model of technical change. This is notoriously a very difficult task to accomplish *per se*, not to say anything to incorporate such a structural model in an econometric general equilibrium model like WARM.

15. Taxes on households and taxes on firms are not distinguished.

16. There is a final item, government subsidies paid to firms, which is kept exogenous in the model. This variable is used for simulation purposes, with the aim of analyzing the effects of policies designed to stimulate technological innovation, export activity, and the like.

17. An important remark concerns the primary factors of production and their international mobility. The demand for labor generated by firms and by the government matches households' labor supply. Thus, the labor market is a national one and this input is internationally immobile. No migration flows are present in the model owing to the paucity of the necessary data. As far as capital is concerned, there is only one type of durable good which is either demanded by households (consumer durables) or by firms (investment goods). Capital is a partly mobile input in the following sense. While durable goods, including capital goods, can be traded internationally, once in place owing to the firm's investment activity, they become perfectly immobile. As already clarified, financial capital flows are not explicitly modeled.

18. The current version of the model does not incorporate a full-blown analysis of the environmental impact of the economic activity. In particular, the model lacks a description of the feedbacks from the environment on to the behavior of the economic agents through households' welfare and firms' cost–benefit analysis. There is however an indirect feedback via the government's reaction functions (for instance, if polluting emissions increase, energy tax rates can be raised).

19. We report the results for these countries as they are the same countries considered in Carraro et al. (1996), of which the present paper can be considered a sequel. Results for the remaining EU countries are available from the authors upon request.

20. This conclusion is consistent with the recommendation contained in the well-knows Delors plan. In Carraro et al. (1996) it is shown that the effects of the fiscal reform on employment can last longer if it is coupled with policies which increase the competitiveness of the labor market or with an income policy which defines the path of wage increases prior to the fiscal reform.

21. The long-run increase in aggregate consumption can be assessed by looking at the consumption/GDP ratio. The increase of this ratio ranges from 0.5% in Spain and The Netherlands to 2.4% in the UK. The increase in consumption is partially offset by a relative reduction of

government expenditure in most countries relative to GDP. In the UK there is also a worsening of the trade balance.

22. Notice that Table 3 contains the difference between the values of employment and GDP in the case where both policy instruments are set by a central government and their values in the case where only the emission tax rates are harmonized.

23. The idea of treating technical change as a unobserved or latent variable is shared by the partial equilibrium approaches of Slade (1989) and Gao (1994), and by the general equilibrium model of Boone et al. (1992).

24. Notice that the firm's investment decision process is a two-stage process. First, the firm decides the aggregate investment expenditure according to the envelope condition of section 3; then, total investment expenditure is split into two components according to the model described here.

25. In the present version of the model, the long-run growth rate of the capital stock is assumed to be constant; $f(X)$ is therefore equal to the growth-rate steady-state value.

26. The weakness of equation (A3) is its *ad hoc*, statistical nature, which contrasts with the theoretical foundations of the rest of the model. However, in order to obtain good econometric results from the latent variable model (a linear filter had to be used), we decided to pay a price in terms of theoretical consistency.

27. Beside m the matrix representation of the system contains the variable vector W and the zeros and ones necessary to reproduce the identities concerning all time-invariant coefficients. Note that the functional specification in (A2)–(A3) has been linearly approximated in (A4)–(A5). The data used here are part of the model database discussed at length in Carraro and Galeotti (1996).

28. The error terms ε and v are assumed to be normally distributed and serially uncorrelated. The covariance matrix of the error terms have been estimated using a maximum-likelihood method. The initial values for the state vector have been estimated using the GLS procedure proposed in Carraro (1985).

29. Notice that our method is close to the one used to decompose a time series into cyclical and seasonal components (see Harvey, 1987).

30. Adding the values generated by equations (A2) and (A3), we obtain the growth rate of total private capital, and therefore the corresponding amount of investment. Given that the amount of investment of each individual firm is determined along the lines discussed in the preceding section, investment expenditures in consumer durables and in housing are determined by the household sector. As clarified in the text, only the first component is endogenized, the second one being obtained as a residual.

Review of International Economics, Special Supplement, 134–147, 1997

Regionalism and the Rest of the World: Theory and Estimates of the Effects of European Integration

L. Alan Winters*

Abstract

There are no satisfactory *ex post* estimates of the effects of regional integration on excluded countries' welfare. Using a formal decomposition of welfare, this paper discusses the factors that might affect these countries' welfare and aspects of their measurement. It then surveys various *ex ante* estimates of the effects of European integration. These suggest that neighboring countries linked tightly to the European economy could lose significantly from the latter's integration, but that for other countries the losses are likely to be very small.

1. Introduction

We have virtually no hard evidence on how regional integration affects economic welfare in the rest of world (RoW). The reasons are that, to date, we have largely failed to measure the effects that theory shows to matter, and that much of what we have measured hardly matters at all. The theoretical and empirical analyses of this important issue are far apart, and the practical difficulties of doing the latter correctly are formidable in terms of both finding data and extracting their messages. Nonetheless, we should try.

This paper starts from the observation—Winters (1997)—that existing *ex post* assessments of the effects of integration are flawed. They look at the RoW's share of the integrating partners' total imports or apparent consumption, and infer welfare conclusions from it, sometimes, wholly inappropriately, invoking the work of Kemp and Wan (1976) in support. Winters (1997) argues that RoW exports are a very poor indicator of RoW welfare, and this paper is intended to explore the possibilities of finding better measures.[1]

Section 2 provides a theoretical basis for quantifying the welfare impacts of integration on RoW, building on the analysis by Baldwin and Venables (1995). It shows that they depend on, *inter alia*, changes in its terms of trade, levels of output and number of firms, its existing trade restrictions, and on induced investment effects. Changes in export and import volumes do matter, but only when interacted with existing trade barriers or traced through to changes in other variables of interest.

To illustrate what is known already, section 3 provides a selective survey of existing quantifications of the effects of European integration on the RoW. Outside the calculations on RoW's exports, there are no *ex post* empirical analyses, so I am thrown back on *ex ante* quantifications based on model simulations. Even then, few direct quantifications exist, so for the most part I try to extract conclusions about RoW from studies

* Winters: International Trade Division, World Bank, Washington, DC 20433, USA. Tel: 202-473-3845; Fax: 522-1159; Email: awinters@worldbank.org. I am grateful to Won Chang for research assistance, to Audrey Kitson-Walters and Minerva Pateña for logistical support, and to Arvind Panagariya, André Sapir, Maurice Schiff, and two referees for comments. The views expressed here are the author's alone. They do not necessarily represent those of the World Bank or any of its member governments.

primarily devoted to European welfare. I restrict myself to the European Union (EU) and its predecessors, but consider several major incidents of integration: the creation of the EEC, the first, second and third enlargements, and the establishment of the Single European Market ("1992"). The conclusion is that integration generally harms the RoW, but not greatly. Section 4 makes a few suggestions for new approaches to the question.

2. A Decomposition of the Welfare Effects of Integration

The Decomposition

Baldwin and Venables (1995) offer a useful decomposition of the welfare effects of regional integration schemes on the integrating partners.[2] In this section, I adapt it to the rest of the world's welfare and propose some minor modifications. I use it to explore the various effects that determine welfare and to discuss their quantification.

Imagine that each country in the world has a representative consumer with indirect utility function $V(p + t, N, E)$, where p is the vector of border prices, t the corresponding vector of tariffs and tariff equivalents, N a matrix of the numbers of (symmetric) varieties of each good available from each country, and E a scalar representing total expenditure on consumption. Expenditure equals income less investment, which, if trade restrictions are the only intervention, implies

$$E = wL + rK + Q'\left[(p + t) - a(w, r, q)\right] + \hat{\alpha}t'm - J, \tag{1}$$

where w is the wage, L the (exogenous) labor supply, r the return to capital, K the capital stock, Q a vector of domestic outputs by commodity, a average production costs by commodity, which depend partly on q, firm-level outputs, $\hat{\alpha}$ a diagonal matrix reporting what proportions of trade costs accrue as rents or revenues domestically, m a vector of net imports, and J a scalar of expenditure on investment.

Taking the total derivative of V and normalizing by the marginal utility of expenditure yields:[3]

$$\begin{aligned}
(dV/V_e) &= (\hat{\alpha}t')dm - m'd(t - t\hat{\alpha}) - m'dp \\
&\quad + (p + t - a)'dQ - Q'\hat{a}_q dq + i'(V_N/V_e) \otimes dNi \\
&\quad + (\bar{r}/\rho - 1)dJ.
\end{aligned} \tag{2}$$

The interpretation of these effects is as follows:

(a) *Perfectly competitive components:*
 (i) $(\hat{\alpha}t')dm$ is the *trade volume effect*: if net imports are subject to a wedge $(\hat{\alpha}t')$, changes in them have first-order effects on welfare.
 (ii) $m'd(t - t\hat{\alpha})$ is the *trade costs effect*: $(t - t\hat{\alpha})$ is the vector of the shares of trade costs lost to the economy; if it changes, the welfare effect depends on the volume of trade affected.
 (iii) $m'dp$ is the *terms of trade effect*: exactly offset by effects in all other countries together.

(b) *Imperfectly competitive effects:*
 (iv) $(p + t - a)'dQ$ is the *output effect*: increases in output generate income/welfare to the extent that consumer prices exceed marginal costs which, we shall assume below, equal average costs.

(v) $Q'\hat{a}_q dq$ is the *scale effect*: increased output per firm (dq) changes average cost by a_q, and this is applied to total output.

(vi) $i'(V_N/V_e)\otimes dNi$ is the *variety effect*: variety, as represented by the number of firms or varieties available, affects welfare directly, possibly by a different amount according to whether the varieties are domestic or foreign.

(c) *Accumulation effects:*

(vii) $(\bar{r}/\rho - 1)dJ$: investment (assumed infinitely lived) generates a flow of welfare determined by the social rate of return ($\bar{r}$), which is discounted back to the present at discount rate ρ; but it entails current foregone consumption (-1).

Perfect Competition

In a competitive world only effects (i)–(iii) are relevant. If the customs union (CU) is small relative to the RoW, its integration will not affect the prices at which the two blocs trade. Thus $dp = 0$, and unless the RoW has tariffs and/or has to change them as a result of the CU, it is quite indifferent to what the partners do. This is an important lesson: *if the RoW is large and pursues free trade, it should not concern itself about integration between small countries.*

If the RoW has constant trade restrictions ($t \neq 0$, $dt = 0$) which generate some revenue ($\alpha > 0$), the trade volume effect will matter, with welfare changes monotonically related to changes in tariff or other revenue. *Integration worsens RoW welfare if and only if it reduces trade tax revenues.* (Recall that for a small CU, $dp = 0$.) It is not strictly possible to say whether changes in revenue will be positive or negative *a priori*. However, if we restrict t to taxes, such that for each commodity i, t_i has the sign of m_i, if integration reduces aggregate trade between CU and RoW, and if tariffs and taxes are not perversely distributed over commodities, there is likely to be a revenue loss. Since tariffs generally outweigh export taxes, however, this is more likely to be related to the RoW's imports than its exports.

The trade cost effect (ii) is important for partners' welfare, for much recent European integration has concerned removing cost-increasing barriers to European commerce; i.e., increasing α. For the RoW, however, it is difficult to envisage changes in the degree of rent-generation from its own barriers arising from other countries' integration.[4]

A more interesting case arises where the CU is large enough to affect the prices of its trade with the RoW (which is also assumed large). Now the RoW has an interest in integration in addition to that stemming from its own trade interventions. The critical transmission mechanism is the terms of trade—item (iii)—and the *RoW benefits (suffers) if the value of its pre-union net imports bundle falls (rises) as a result of integration.* Since reductions in CU imports from the RoW reduce the prices of these imports, there is an (indirect) link from RoW exports to RoW welfare, but it is only half the story.

An early, but subsequently strangely neglected, exploration of the terms of trade effect was by Mundell (1964). In a three-country three-good model, Mundell argued that, given gross substitutability, a union between countries A and B would increase aggregate demand for A's and B's goods relative to that for C's (RoW's) via trade diversion. This would need to be offset by price changes which would reduce the price of C's good relative to an average of A's and B's—i.e., it would worsen C's terms of trade, $m'dp > 0$.

Endogenous Trade Policy

The improvement in the union's terms of trade is automatic in Mundell's analysis, but it could also stem from policy changes in the union.[5] Viner (1950) observed that a historically important argument for the creation of a customs union is to increase market power. Arndt (1968) showed that integration would increase the optimum tariff of the union partners, and Bond et al. (1996) show that reducing barriers within an existing customs union will make the latter more aggressive in the sense of raising its optimum tariff against the RoW. Both of these observations imply that, if the union applies its optimum tariffs, the RoW will suffer as the terms of trade are turned further against it. Political economy considerations could mitigate this conclusion, if "preference dilution" reduces the power of national lobbies to obtain protection from union authorities relative to what they had *vis-à-vis* national authorities (de Melo et al., 1993). Winters (1993, 1994) suggests, however, that dilution is far from inevitable and that the internal decision processes of the union could also foster protectionism. Thus changes in union trade policy seem more likely to enhance than to offset Mundell's adverse terms of trade effects for the RoW.

   If we endogenize the union's trade policy we should do the same for the RoW. This introduces further complex determinants of dp in equation (2) related to terms in dm and dt. Since the union affects the partners' net import demand and their strategic behavior—they now react as one not two separate entities—it also affects the RoW's optimum tariff. In general one would expect integration to increase the partners' elasticities of excess demand and the jointness of the partners' retaliation to make RoW aggression less attractive. Both factors will tend to reduce the RoW's optimum tariffs.[6] Relative to leaving its tariff unchanged in the face of integration, this cut will reduce the welfare loss the RoW feels, because although the terms of trade turn further against it, there are offsetting benefits from increased trade.

   One caveat to the discussion above is to recognize that the RoW is not a single entity. If, instead, it comprised myriad small countries each of which traded with many partners and which could not coordinate their trade policies, all talk of optimum tariffs would be moot. European integration may well worsen their terms of trade and may increase the European partners' market power, but no mitigation would be feasible for any member of the RoW. Even if, in the real world, some members of the RoW do have countervailing market power, the majority will not, and even those that do would probably wish to maintain nondiscriminatory policies, so that they will be further restrained in their response to European integration by the damage that it will do to their other trade flows.

Imperfect Competition

The second line of equation (2) contains three effects pertaining to imperfect competition: an output effect, a scale effect and a variety effect. Different approaches to modeling imperfect competition stress different terms, but all present pretty formidable measurement problems.

   The output effect, (iv), reflects the fact that if the cost of a unit of output differs from its price, producers and consumers will value those units differently and, in equilibrium, the economy will produce nonoptimal amounts of the various goods. Under these circumstances policies that affect output will have direct first-order effects on welfare. Several points are worth noting about this term.

First, Baldwin and Venables' assumption that average costs, *a*, depend on firm output (*q*) but not total output (*Q*) implies that industry expansion occurs only through the replication of firms, so that industry average and marginal costs are equal.

Second, the output effect refers to changes in total output by industry, not trade. Even if the RoW increases its sales to the union as a result of integration, welfare will not rise if the exports are diverted from domestic consumption—if, for example, there are capacity constraints. Moreover, if one industry is to expand, it is likely that another will contract (possibly squeezed by increased imports), and what then matters is the difference between the price–cost margins of the two industries.

Third, the price–cost margin applied to changes in output should include allowance for normal profits and should, strictly, refer to social costs not accounting costs. The latter point could be relevant in developing countries, where there are frequently tax/subsidy distortions: a decline in the output of a subsidized product could be beneficial. It will also be relevant if product market imperfections are partly captured by workers in the form of higher wages: shifting output towards products which pay rents to workers is desirable (Dixit, 1988).

As expressed, the scale effect, (v), refers to internal economies of scale. The term $\hat{a}_q dq$ is the change in average costs arising from integration: here these are due only to increases in firm level output, but in any practical *ex post* measurement, changes in costs arising from any cause—e.g., external economies, technology transfer, etc.—would be included. The presence of *Q* in the term reflects the fact that changes in average cost apply to total output, not just to trade.

The critical factors behind the welfare effect on the RoW are (1) the extent to which its costs are affected by integration, which is likely to be larger the greater an industry's dependence on the union markets, and (2) the greater the volume of output to which cost changes are applied. If *a*(*q*) has locally a constant elasticity of costs with respect to output ($\gamma < 0$), and if we consider only the impact effect of changes in RoW exports to the CU, the scale effect for a single firm becomes $\gamma a dx$, which, with the minus sign in equation (2), would reduce welfare if exports fell.[7] This goes a little way towards rehabilitating the focus on the RoW's exports to the union, but only a little: RoW imports might also be impacted and affect the scale of operations for local industry; the sensitivity of costs to output is critical; and cost changes will affect local sales, so the focus only on impact effects (i.e., assuming *dq* = *dx*) is inappropriate.

The variety effect, (vi), is perfectly straightforward conceptually but nearly impossible to measure practically. Variety confers benefits on both consumers and users of intermediate inputs, and Romer (1994) argues that enlarging the set of goods available is probably the major benefit of international trade. I extend Baldwin and Venables by disaggregating the numbers of goods by source, because in most circumstances the addition of an extra good will have different welfare effects according to how many similar goods exist and the amounts of them that are consumed. These latter factors can differ across suppliers.

In practical terms it is impossible to measure how many varieties are available to consumers and virtually impossible to measure the number of firms offering goods for sale, which in simple models is assumed to be the same thing. One might argue (e.g., Krugman, 1989) that variety is proportional to GDP and thus that the effects of integration must balance the likely positive effects of increasing the set of goods produced by the union against the decrease, if any, in the RoW. Even this, however, begs the question of determining the value of variety, V_N. Overall, it is difficult to see how such effects can be quantified in broad macroeconomic studies of actual integration, but they are amenable to *ex ante* modeling and they may be tractable in detailed sectoral studies. Although changes in the value of the RoW's exports to the union will have some impact on variety, it is very difficult to believe that any practical quantification can be based solely on that link.

The three imperfectly competitive effects interact in quite complex ways. For example, if variety effects are strong, firm size effects are likely to be small. Baldwin and Venables explore these connections for partner welfare, and Winters (1995) for nonmembers.

Accumulation Effects

The final term of equation (2) captures the medium-term pseudodynamics of integration along the lines of Baldwin (1989, 1992).[8] These focus on the incentives created by integration to change capital stocks. As noted above, the welfare benefits relate to the difference between the social returns to investment and its cost. Assuming that the additional capital stock is infinitely-lived, increments generate a flow of welfare of $\bar{r}dJ$ per annum, which has a net present value of $(\bar{r}/\rho)dJ$; against this dJ units of current welfare must be sacrificed to make the investment.

The effects of integration on investment have attracted much comment recently, frequently being quoted as the principal source of benefit of schemes such as NAFTA (e.g., Young and Romero, 1994). Strong reservations have been expressed by non-members that they have lost investment to the integrating bloc—either through the redirection of foreign investments, say to Mexico from elsewhere in Central America, or through domestic capital outflow as, say, firms from EFTA boosted investment in the EU as the single market was created. Equation (2) offers no hint of the mechanisms involved here and I shall not discuss them, but I have a few comments on their consequences.

First, if one has a notion of an *anti-monde* for investment, this element of equation (2) may potentially be operationalized. Investment is measured fairly directly and estimates of rates of return and of time discounts are available. Second, there is a welfare effect *only* to the extent that social returns ($\bar{r}$) exceed the rate of time discount. This is possible for a number of well-known reasons, but it cannot be taken for granted.

Third, the presumption of equation (2) is that investment is domestically financed: the whole return ($\bar{r}$) and the whole cost (-1) enter RoW's welfare measure. Thus it copes more easily with Baldwin's model of changes in the incentives for local investment than with cross-border flows. If the RoW diverts investment to the CU, its local investment is reduced ($dJ < 0$), but no allowance is made in (2) for the return flow of repatriated profits. One could capture this by treating it as part of $\bar{r}$ for the diverted flow, but would then require separate terms in $\bar{r}_D \, dJ_D$ for the lost RoW domestic investment, and $\bar{r}_F \, dJ_F$ for the increased outflow ($dJ_D = -dJ_F < 0$). Although private rates of return on foreign and domestic investment would be equalized, it is possible that spillovers and taxes will make domestic investment more valuable—i.e., $\bar{r}_D > \bar{r}_F$. This becomes even more important when we consider the RoW's capital inflows from the CU, because then the investment costs no local resources, but may still generate a social return to the extent that $\bar{r}_D$ exceeds the repatriated income flow.

3. Quantitative Estimates

I turn now to a selection of quantitative estimates. I restrict attention to studies which deal with the direct effects of European integration, are quantitative, and are based on an explicit model. In the light of the comments above, I consider only the few studies which have attempted to move beyond the examination of trade shares.[9] The studies I survey are primarily concerned with the effects of integration on the European partners themselves. Thus the results for the RoW are usually incomplete (often only trade and trade prices) and implicit. They typically have received very little attention from their authors, and so should not be considered in any sense definitive.

The Methodology for Quantifying Integration Effects

There is an important distinction between *ex ante* and *ex post* approaches to quantifying the effects of integration. *Ex ante* estimates are typically made before the event to estimate its likely impact, but the most useful definition is that, regardless of the date of the study, *ex ante* models make no use of data pertaining to the post-integration period. *Ex ante* quantifications are necessarily model-based: they rely on formulating a view of how the economy will respond to the policy changes required by integration and then implementing those changes. They are typically long on theory and, at least in recent years, calibrated rather than estimated because the theoretical complexity of the models puts them far beyond the reach of what can be extracted from current datasets.

A critical component of any assessment of integration is the *anti-monde*, the alternative situation if integration had not occurred. In *ex ante* models this is usually taken as the actual situation in the base year—the year of the data to which the model is calibrated. Thus, for example, Haaland and Norman's (1992) results are as if "1992" had occurred in 1985. This is clear conceptually and avoids the problems of projecting the economy into the future to when "1992" is completed. It means, however, that the results are illustrative thought experiments, not unconditional predictions.

Ex post studies, as the terminology implies, come after integration and rely on post-integration data to reveal integration effects. In these cases, the real data describe the "with integration" situation, and the challenge is to construct the "without integration" *anti-monde*. This is usually defined—perhaps only implicitly—as no change in some variable of which researchers compare "before" and "after" values. Many such studies look—inappropriately, for assessing welfare effects on the RoW—at the evolution of the RoW's share of the partners' imports across a period of integration. This is very straightforward practically: trade data are readily available and fairly accurate; trade shares move slowly and fairly predicably and so can be constructed for an *anti-monde*; and as a matter of positive economics, the RoW's share of partners' imports is likely to be affected by integration to a greater proportionate extent than almost any other variable. Given that *ex post* research relies on event analysis—recognizing deviations in a variable that are roughly contemporary with integration but cannot be explained any other way—the attractions of import shares are obvious.

The variables identified in section 2 as determining RoW welfare are much less promising empirically. Even the major European integrations will have only limited proportionate effects on variables such as world prices and RoW output and scales of operation. Moreover, the other factors that influence these variables are more difficult to model, and hence to allow for in defining an *anti-monde*. Some of the variables—for example, the extent of variety or the degree of economies of scale—more or less defy measurement. Thus the scope for identifying integration shocks in the variables critical for RoW welfare seems, *prima facie*, very limited. Indeed, so limited that I have been unable to identify a single *ex post* study using the concepts of section 2.

The Effects of European Integration: A Selective Survey

I turn now to a selection of the studies themselves, organizing them along the same lines as section 2. As *ex ante* studies, they are more representative of what economists believe the effects of integration to be than of what those effects have actually been.

Perfect competition   Petith (1977) implemented Mundell's (1964) model of the terms of trade. He derived expressions for the demand for imports in terms of prices and elasticities and then shocked these by removing tariffs on intrabloc trade. He used elasticities of substitution based on other studies and thus essentially undertook a calibration study. Because he assumed no distortions in the RoW, he effectively considered only the finite change analog of the term $m'dp$ from equation (2). Petith did not compare his results with actual data, perhaps because his schematic three-country model was so unrealistic geographically.

Petith predicted large terms-of-trade effects in each of his three experiments. The partners' terms of trade generally improved by 2–3%, generating welfare benefits of 0.3–0.9% of their GDP. These, of course, were matched by corresponding losses to the RoW. Sapir (1992) has argued that Petith's results are flawed because they assume that average European tariffs were the same before and after integration, ignoring, he says, the significant cuts in external tariffs introduced in the Dillon and Kennedy Rounds. This may be seen as a problem of defining an *anti-monde*. Petith is essentially looking at, say, 1960 and asking, *in that world*, what happened when the Europeans offered each other preferential tariff reductions. Sapir, on the other hand, might be saying that since by, say, 1970 the world economy would have lowered tariffs anyway, it makes sense to ask what was the effect of the preferential reductions in the context of a "low tariff" world.

Sapir goes on to argue that the Kennedy and Dillon Rounds stemmed at least partly from the process of European integration (p. 1,500) because they represented the US response to that event. This is possible, although—see Winters (1993)—it suggests strong practical evidence that the RoW believed it would lose from European integration. By seeking a negotiation in response to integration, the United States was signaling that integration had increased its returns to negotiation. Since it is difficult to see how the absolute level of US welfare would be higher if it had negotiated liberalization with "the Six" jointly rather than separately, the extra incentive seems more likely to have stemmed from a perception that integration would worsen the US situation if no liberalization were agreed. Sapir is essentially arguing that, even though, *ceteris paribus*, its impact may have been harmful, European integration actually benefited the RoW because it initiated a positive interaction *in policies*. In terms of (2), Sapir is arguing that Petith overestimates the losses in dp and ignores (i), (ii), and a term in tariff changes that would be needed for finite dt.

A second study dealing directly with terms-of-trade effects is Kreinin and Plummer (1992). Examining the Southern Enlargement of the EC, these authors identified major export flows from ASEAN and South Korea to the EC-9 for which Spain, Portugal, and Greece provided the main competition. For each of these goods they assumed that removing the tariff on the accedants' sales would reduce the EC-9 internal price by two-thirds of the tariff and that, as small suppliers, ASEAN and Korea would suffer the same fall in price on their exports to those markets. Assuming no corresponding quantity adjustment, the terms-of-trade loss to the exporters is just $m'dp$, the initial quantity multiplied by this change in price. Taking all the relevant products, Kreinin and Plummer identified losses of $468 million for ASEAN and $324 million for Korea. Exports to the southern accedants were treated similarly except that their internal prices were assumed to fall by the full extent of the tariff removed from EC-9 supplies. Losses in these markets amounted to about $40 million for each supplier.

Pomfret (1993) argued that Kreinin and Plummer had overestimated the welfare effect. First, the assumed changes in the domestic price were too large: Kreinin and Plummer assumed that the partners could expand sales significantly with little or no increase in costs and that EC domestic suppliers did not ameliorate the downward pressure on prices by cutting back their output. Second, Kreinin and Plummer assumed that the exporters made no quantity adjustment. If factors of production could have moved out of the affected sector, or exports could have been diverted to other markets, the quantity of exports to the EC would have fallen (exacerbating the loss of export revenue *from that commodity*), but the welfare cost would have been lower because the factors would have taken a smaller cut in their marginal revenue productivity. This is essentially asking for a finite change version of equation (2).

There seems much to be said for Pomfret's critique, but this should not obscure the virtue of Kreinin and Plummer's approach in general. Their focus is on the right question and one which, at least in principle, is amenable to subsequent *ex post* analysis.

A further multicommodity partial equilibrium estimate of integration effects is Cawley and Davenport (1987) on the effects of "1992"—completing the single market. For each of six EC markets they constructed simple models of many sectors identifying three suppliers—domestic, rest of EC, and RoW. They then explored the effects of (I) the reduction in barriers to imports from the rest of EC plus smaller reductions on imports from the RoW, and (II) the reduction of domestic regulations affecting domestic and rest of EC production costs. All suppliers displayed upward-sloping supply curves; hence because integration tended to reduce sales by the RoW, it also drove down their prices, i.e., $dp_i < 0$ for RoW's exports.

In exercise I, which concerned the reduction of trading costs, the RoW experiences losses of exports of 2–2.5%, which translate into price falls of 0.3–0.5% given the assumed elasticities of supply of 5 and 6. Stage II, referring to the relaxation of cost-increasing EC regulations, has greater effects, with RoW exports falling by 5.7–7.7% which translates into price declines of 1.5–1.8%. Overall, therefore, the RoW is expected to experience about a 2% decline in export prices as a result of "1992" before any allowance is made for induced changes in EC market structure or economies of scale. Since the latter were slated to increase the gains to the EC by a factor of two or more, the overall effects on the RoW were likely to be significantly larger. On the other hand, the calculations reported here also make no allowance for induced declines in the prices of EC exports (i.e., RoW imports) so they are not full terms-of-trade effects.

Davenport and Page (1991) build on Cawley and Davenport's results with the object of isolating the effects of "1992" on developing countries. They refine the estimate of $m'dp$, but, because they have no model of the RoW, cannot explore any of the other effects in equation (2). They disaggregate Cawley and Davenport's RoW, refine their estimates of trade barriers, and assume that "1992" boosts EC GDP by 5%. The last effect helps to offset the diversionary consequences of preferences and permits developing countries, especially primary producers, to expand their sales and raise their prices (to all buyers). Under these assumptions, manufactured exports are roughly unchanged by "1992," while primaries show significant growth and terms-of-trade benefits. This growth effect is clearly an important component of the welfare effects of deepening EC integration when measured relative to a "no further integration" *anti-monde*. If the *anti-monde* were also growth-enhancing, however—e.g., multilateral liberalization or domestic deregulation—it would not be significant.

Davenport and Page also take account of the beneficial effects on developing countries' import prices of the EC's enhanced efficiency as a result of "1992." Greater competition and economies of scale drive down EC manufactures' prices and thus EC export prices, which further improves the RoW's terms of trade. Combining the effects on import and export prices, Davenport and Page estimate an overall terms of trade *gain* of 2% of developing countries' exports to the EC.

Imperfect competition   The recognition of internal economies of scale requires researchers to model industries in an imperfectly competitive fashion. Norman (1989) does this in a way that offers explicit welfare estimates for the RoW. He considers two sectors—motor vehicles and pharmaceuticals—and conducts a variety of integration experiments. He finds that excluded countries always lose from integration. The size of the effects is instructive. When an excluded country sells a significant share of its total output to the integrating bloc, it can suffer significantly—as Sweden does from "1992" in Norman's results. To the extent that it is a consumer of EC goods, on the other hand, a country gains from declines in the prices of EC exports. The former effect tends to dominate, however, because price–cost margins considerably exceed the falls in EC prices, so that a country needs a large trade deficit with the EC in order to show net benefits—e.g., Norway in motor vehicles. These results also include the effects of Norway and Sweden's own barriers ($t'dm$). The RoW outside Scandinavia—which figures only as a supplier—suffers net losses from integration in both sectors, but they are very small, because it sells only small fraction of its output in the EC.

Haaland and Norman (1992) reiterate these findings in a fully closed computable general equilibrium model with endogenous factor returns. The RoW, now represented as Japan and the USA, loses slightly from integration in Europe, but because of the small proportion of their transactions that are with the EU, not by much. EFTA, on the other hand, sold around 20% of its GDP in the EC during the mid-1980s. Thus it was predicted to experience much greater costs from exclusion and greater benefits from accession.

Haaland and Norman's results illustrate how comparative advantage and factor rewards are affected by integration. The benefits of "1992" stem mainly from the pro-competitive effects of increased market integration, which reduce firms' market power. (Firms are less able to maintain local fiefdoms in which they can exercise market power to keep prices up.) This is more significant the less competitive the industry was initially, and so "1992" has stronger efficiency effects in the EC in more oligopolistic industries. This, in turn, implies that "1992" moves EFTA's comparative advantage away from industries such as engineering, where the EC makes gains, towards processing primary products, where it does not. It is precisely because engineering has larger margins—$(p + t - a)$ in equation (2)—than primary processing that EFTA suffers a fall in income from "1992."

Gasiorek et al.'s (1992) model is similar to Haaland and Norman's, and it appears to produce similar results. They report two decompositions of the welfare effects of "1992" based quite closely on equation (2), but, unfortunately, only for the EC. Although one can infer some of the consequences for the RoW, one cannot do the job completely. Gasiorek et al. identify substantial reductions in EC imports from the RoW. They find, however, that the associated changes in prices, while negative, are rather small, because the squeeze on RoW margins in the EC, as local industry benefits from reduced trading costs and greater competition, is partly offset by increases in cost as the RoW firms slide *up* their average cost curves. The net effect is relatively small effects on the import side of the EC's terms of trade; the terms-of-trade effect is, of course, equal and opposite for the RoW, but the increased costs in the RoW reduce its welfare via the average cost effect, (v) in equation (2). Unfortunately, Gasiorek et al. do not report this magnitude for RoW.

Gasiorek et al. also report EC export prices to the RoW. These fall significantly as EC firms exploit their newfound efficiency, and the net effect is a shift in the EC–RoW terms of trade in the latter's favor. There is no doubt that these falls in EC export prices and related increases in exports are directly beneficial to the RoW, but we do not know how they interact with the RoW's own trade distortions, nor the extent to which the associated changes in RoW scale and production costs offset them. As in Haaland and Norman, however, the most likely conclusion is that, outside the heavily EC-dependent countries of Europe and the Mediterranean basin, the effects are proportionately rather small.

Accumulation   The general equilibrium models opened up, for the first time, the possibility of identifying the effects of integration on factor rewards, which, in turn, raised questions about the accumulation of capital. Baldwin (1989, 1993) provided the theory and orders of magnitude, and these effects are now beginning to be quantified by integration modelers, such as Mercenier and Akitoby (1993) and Harrison et al. (1994). Accumulation typically magnifies the output effects of integration, but as equation (2) shows, it affects welfare only if (initial) returns differ from the rate of time discount.

Mercenier and Akitoby's results suggest that allowing for accumulation turns the 1992 terms-of-trade shock on the RoW from negative to positive. This presumably reflects the fact that increased EC investment increases its excess supply of exportables, while the opposite happens in the RoW. Harrison et al. (1994) do not report accumulation for the RoW, but their static results suggest that the rate of return on capital in RoW increases as a result of "1992." Given greater increases inside the EC, however, we should probably anticipate the same terms of trade improvement as Mercenier and Akitoby found.

4. Conclusion

The correct estimation of the effects of integration on the rest of the world is a major challenge, requiring the assessment of variables such as the terms of trade, scale of production, other distortions, and the extent of variety. To date, no *ex post* study has attempted the task—all we have are studies of one small and indirect part of it. A number of *ex ante* studies do exist, however, which suggest that integration harms the rest of the world. Except where a country is highly dependent on the integrating bloc, however, the effects are likely to be small relative to GDP.

The need to address the RoW effects of integration is pressing, and so I conclude with a few comments on how we might proceed. Winters and Chang (1997) have initiated a study of movements in relative partner and nonpartner import prices into Spain following its accession to the EU in 1986. They confirm the substantial difficulties of events studies on noisy data such as on prices and have to impose substantial structure on the data in order to derive results. The latter are encouraging, however, since they indicate some evidence of differences in the evolution of prices. A second approach is to supplement studies of the RoW's exports with those of its imports. In

Winters (1985) I found that UK-manufactured exports to some nonpartner markets were lower following UK accession to the EC than would have been expected otherwise. If the *anti-monde* behind this finding is appropriate, it suggests RoW welfare losses.

A third approach is detailed case studies of integration looking at prices, product availability, and investment following an integration. Firm or sectoral level data may reveal changes in either EU or RoW firms' exports, from which conclusions about scale and variety could be derived. These would, of course, not be immediately generalizable, but given our present ignorance any information would be useful.

None of these exercises will reflect all the factors noted above, but given that regional integration strikes at one of the pillars of the multilateral trading system, it seems important to try to quantify its effects empirically. Only in that way can the fears of excluded countries be legitimately justified or set aside, and the multilateral rules for regional integration rationally assessed.

References

Anderson, Kym and Richard Blackhurst, *Regional Integration and the Global Trading System*, Hemel Hempstead: Harvester Wheatsheaf, St Martins Press, 1993.

Anderson, Kym and Hege Norheim, "History, Geography and Regional Economic Integration," ch. 2 in Kym Anderson and Richard Blackhurst (eds.), 1993, pp. 19–51.

Arndt, Sven W., "On Discriminatory vs. Non-preferential Tariff Policies," *Economic Journal* 78 (1968):971–9.

Baldwin, Richard E., "The Growth Effects of 1992," *Economic Policy* 9 (1989):247–82.

———, "Measurable Dynamic Gains to Trade," *Journal of Political Economy* 100 (1992):162–74.

Baldwin, Richard E. and Anthony J. Venables, "Regional Economic Integration," in Gene M. Grossman and Kenneth Rogoff (eds.), *Handbook of International Economics*, vol. 3, Amsterdam: North-Holland, 1995.

Bond, Eric W., Costantinos Syropoulos, and L. Alan Winters, "Deepening of Regional Integration and Multilateral Trade Agreements," CEPR discussion paper 1317, 1996.

Cawley, Richard and Michael Davenport, "Partial Equilibrium Calculations of the Impact of Internal Market Barriers in the European Community," economic paper 73, European Commission, Brussels, 1988.

Davenport, Michael and Sheila A.B. Page, *Europe: 1992 and the Developing World*, London: Overseas Development Institute, 1991.

De Melo, Jaime and Arvind Panagariya, *New Dimensions in Regional Integration*, Cambridge: Cambridge University Press, 1993.

De Melo, Jaime, Arvind Panagariya and Dani Rodrik, "The New Regionalism: A Country Perspective," ch. 5 in De Melo and Panagariya (eds.), 1993, 159–99.

Dixit, Avinash K., "Optimal Trade and Industrial Policies for the US Automobile Industry," ch. 6 in Robert Feenstra (ed.), 1988, pp. 141–65.

Feenstra, Robert C., *Empirical Methods of International Trade*, Cambridge: MIT Press, 1988.

Francois, Joseph F. and Clinton R. Shiells, *Modelling Trade Policy: Applied General Equilibrium Assessments of North America Free Trade*, Cambridge: Cambridge University Press, 1994.

Gasiorek, Michael, Alasdair Smith, and Anthony J. Venables, "'1992': Trade and Welfare—A General Equilibrium Model," ch. 2 in L. Alan Winters (ed.), 1992, pp. 35–61.

Haaland, Jan I. and Victor D. Norman, "Global Production Effects of European Integration," ch. 3 in L. Alan Winters (ed.), 1992, pp. 67–89.

Harrison, Glenn W., Thomas F. Rutherford, and David G. Tarr, "Product Standards, Imperfect Competition, and the Completion of the Market in the European Union," World Bank Policy Research working paper 1293, Washington, DC, 1994.

Kemp, Murray C. and Henry Y. Wan, "An Elementary Proposition Concerning the Formation of Customs Unions," *Journal of International Economics* 6 (1976):95–7.

Kowalczyk, Carsten, "Paradoxes in Integration Theory," *Open Economies Review* 3 (1992):51–9.

Kreinin, Mordechai, E. and Michael G. Plummer, "Effects of Economic Integration in Industrial Countries on ASEAN and the Asian NIEs," *World Development* 20 (1992):1345–66.

Krugman, Paul R., "Differences in Income Elasticities and Trends in Real Exchange Rates," *European Economic Review* 33 (1989):1031–54.

Lloyd, Peter J., "Regionalisation and World Trade," *OECD Economic Studies* 18 (1992):7–43.

Mercenier, Jean and Bernardin Akitoby, "On Intertemporal General Equilibrium Effects of Europe's Move to a Single Market," Institute of Empirical Macro Economics discussion paper 87, 1993.

Mundell, Robert A., "Tariff Preferences and the Terms of Trade," *Manchester School Economic Social Studies* 32 (1964):1–13.

Norman, Victor, "EFTA and the Internal European Market," *Economic Policy* 9 (1989):423–66.

Petith, Howard C., "European Integration and the Terms of Trade," *The Economic Journal* 87 (1977):262–72.

Pomfret, Richard, "Measuring the Effects of Economic Integration on Third Countries: A Comment on Kreinin and Plummer", *World Development* 21 (1993):1435–7.

Romer, Paul, "New Goods, Old Theory and the Welfare Costs of Trade Restrictions," *Journal of Development Economics* 43 (1994):5–38.

Sapir, André, "Regional Integration in Europe," *Economic Journal* 102 (1992):1491–506.

Srinivasan, T. N., John Whalley, and Ian Wooton, "Measuring the Effects of Regionalism on Trade and Welfare," ch. 3 in Kym Anderson and Richard Blackhurst (eds.), 1993, pp. 52–79.

Viner, Jacob, *The Customs Union Issue*, New York: Carnegie Endowment for International Peace, 1950.

Winters, L. Alan, "Separability and the Modelling of UK Exports to Five Countries," *European Economic Review* 27 (1985):335–53.

———, *Trade Flows and Trade Policy After 1992*, Cambridge: Cambridge University Press, 1992.

———, "The European Community: A Case of Successful Integration?" ch. 7 in De Melo and Panagariya (eds.), 1993, pp. 202–27.

———, "The EC and Protection: The Political Economy," *European Economic Review* 38 (1994):596–603.

———, "European Integration and Economic Welfare in the Rest of the World," presented at the Conference on European Economic Integration, Nantes, June 1995.

———, "Regionalism vs. Multilateralism," Centre for Economic Policy Research discussion paper 1525, London, 1996.

———, "Regionalism and the Rest of the World: The Irrelevance of the Kemp–Wan Theorem," *Oxford Economic Papers* 49 (1997):228–34.

Winters, L. Alan and Won Chang, "Regional Integration and Prices of Imports: An Empirical Investigation," Policy Research Working Paper, 1782.

Wooton, Ian, "Preferential Trading Arrangements: An Investigation," *Journal of International Economics* 21 (1987):81–97.

Young, Leslie and José Romero, "A Dynamic Dual Model of the North American Free Trade Agreement," ch. 10 in Francois and Shiells (eds.), 1994, pp. 301–27.

Notes

1. Winters (1995) offers an extended discussion of the issues raised in this paper.
2. Other authors offering similar decompositions include Wooton (1986), Kowalczyk (1992), and Harrison et al. (1993). Lloyd (1992) and Srinivasan et al. (1993) are among those who consider the RoW in a way related to this paper.

3. Generalizing to finite charges is conceptually feasible, even if practically complex (Harrison et al., 1993).

4. Integration may affect the distribution of rents from the CU's VERs between importer and exporter firms, but these changes will be reflected in the RoW's terms of trade, p, not its parameter α.

5. Winters (1996) offers a broad survey of the effects of regional arrangements on trade policies.

6. Bond et al. (1996) support this view in a one-shot tariff negotiation, but show that, in a repeated game setting in which low tariff equilibria are supported by trigger strategies, other outcomes are possible.

7. $a_q = \gamma a/q$; $dq = dx$; and for a single firm we use q instead of Q.

8. Genuine dynamics do not appear in (2), for, in truth, we know too little to model, let alone measure, them.

9. Within the trade-share studies, Sapir (1992) and Anderson and Norheim (1993) are among the most useful.